DEATH: BEYOND WHOLE-BRAIN CRITERIA

PHILOSOPHY AND MEDICINE

Editors:

H. TRISTRAM ENGELHARDT, JR.

Center for Ethics, Medicine, and Public Issues,
Baylor College of Medicine, Houston, Texas, U.S.A.

STUART F. SPICKER

School of Medicine, University of Connecticut Health Center,
Farmington, Connecticut, U.S.A.

VOLUME 31

DEATH: BEYOND WHOLE-BRAIN CRITERIA

Edited by

RICHARD M. ZANER

Vanderbilt University, Nashville, Tennessee, U.S.A.

KLUWER ACADEMIC PUBLISHERS

DORDRECHT / BOSTON / LONDON

Library of Congress Cataloging-in-Publication Data

Death : beyond whole-brain criteria / edited by Richard M. Zaner.
        p.   cm. — (Philosophy and medicine ; v. 31)
  Based on a symposium held on Oct. 18–20, 1984, at Vanderbilt
University in Nashville, Tenn.
    Includes bibliographies and index.
    ISBN 1–55608–053–0 (U.S.)
      1. Death—Proof and certification—Congresses. 2. Brain death—
Congresses. I. Zaner, Richard M. II. Series.
    [DNLM: 1. Brain Death—congresses. 2. Death—congresses.
3. Ethics, Medical—congresses. W3 PH609 v. 31 / W 820 D2853 1984]
RA405.A1D43 1988
616.07'8—dc 19
DNLM/DLC
for Library of Congress   87–28491

Published by D. Reidel Publishing Company
P.O. Box 17, 3300 AA Dordrecht, The Netherlands

Kluwer Academic Publishers incorporates
the publishing programmes of D. Reidel, Martinus Nijhoff,
Dr W. Junk and MTP Press.

Sold and distributed in the U.S.A. and Canada
by Kluwer Academic Publishers,
101 Philip Drive, Norwell, MA 02061, U.S.A.

In all other countries, sold and distributed
by Kluwer Academic Publishers Group,
P.O. Box 322, 3300 AH Dordrecht, The Netherlands

Printed in the Netherlands

TABLE OF CONTENTS

v

PART IV / THE CULTURAL CONTEXT 217

EDITOR'S PREFACE

From the tone of the report by the President's Commission for the Study of Ethical Problems in Medicine and Biomedical and Behavioral Research, one might conclude that the whole-brain-oriented definition of death is now firmly established as an enduring element of public policy. In that report, *Defining Death: Medical, Legal and Ethical Issues in the Determination of Death*, the President's Commission forwarded a uniform determination of death act, which laid heavy accent on the significance of the brain stem in determining whether an individual is alive or dead:

An individual who has sustained either (1) irreversible cessation of circulatory and respiratory functions, or (2) irreversible cessation of all functions of the entire brain, including the brain stem, is dead. A determination of death must be made in accordance with accepted medical standards ([1], p. 2).

The plausibility of these criteria is undermined as soon as one confronts the question of the level of treatment that ought to be provided to human bodies that have permanently lost consciousness but whose brain stems are still functioning. Are such entities persons, patients to whom continued medical treatment is due? Or have the individuals died, leaving a residual body that continues to respire spontaneously because of the presence of an intact brain stem? As is noted by the contributors to this volume, this problem was subsequently addressed by the President's Commission when in its report, *Deciding to Forego Life-Sustaining Treatment*, it recommended that all medical treatment could be discontinued when patients were permanently unconscious, given the approval of the patient's family ([2], p. 6). This view has subsequently been sustained by the Supreme Judicial Court of Massachusetts, which reviewed the case of the late Paul Brophy, whose treatment was subsequently stopped at the request of the family, following the previously stated wishes of Paul Brophy, who had lost all consciousness (Patricia E. Brophy *v*. New England Sinai Hospital, Inc., No. 4152 [Sup. Judicial Ct., 11 September 1986]). Such practical decisions, judged sensible or not, leave many central theoretical questions unanswered, for they put unconscious human bodies in a strange limbo where we are often uncertain whether they are or are not the bodies of dead persons.

Such questions which concern the ontological status of human bodies, where there is no further possibility of consciousness due to irrevocable and irreversible destruction of the higher brain centers, led to a symposium on October 18–20, 1984, at Vanderbilt University in Nashville, Tennessee: "When Are You Dead?: Critical Appraisals of Whole-Brain and Neocortical Definitions of Death." Presentations made at that symposium provided the major structure for this volume. Subsequent discussions and reflections have given it its final form. We wish to express our deep gratitude to the authors for working with us over this period of time to provide a critical reexamination of the philosophical and public policy assumptions involved in our current understanding of the various definitions of death.

We take this opportunity to express our appreciation to the Tennessee Committee for the Humanities, a not-for-profit State committee of the National Endowment for the Humanities (Grant #A7-24041-21), and the Endowment's Division of Research Programs for its additional material support (Grant #RD-20514). St. Thomas Hospital, Mr. Nelson Andrews, the Tennessee Bar Association, Vanderbilt University School of Medicine, Vanderbilt University Hospital Nursing Continuing Education/Quality Assurance, Vanderbilt University Divinity School, Vanderbilt University Department of Philosophy, Vanderbilt University Medical Center (Division of Continuing Education), and the Center for Clinical and Research Ethics provided additional and generous support to conduct the symposium that led to this volume. We wish also to thank the many individuals who contributed their time and energies to the success of this program: John S. Derryberry, Gerald M. Fenichel, A. Everette James, John Post, Roscoe R. Robinson, and Richard S. Stein. Many others have subsequently aided in the preparation of the manuscript for this volume. Here we are pleased to acknowledge with gratitude Mary Ann Gardell Cutter and Dixie Hargraves.

This volume provides a basis for the further critical examination of what it means to cease to be a person in this world. Though as a philosophical problem this was always available for us to consider, as this volume reveals, modern biomedical science and modern biotechnology have added impetus to such discussions. We now understand more clearly how not only the brain but the higher centers of the brain are necessary for conscious personal life. The new technologies now make it possible for us to sustain organic bodies which will never again embody a conscious person. As the contributors to this volume have shown, it is now necessary to reexamine what it means to be a person alive in this world.

It might also be mentioned that these essays contribute importantly to the ongoing discussions about organ transplantation and donation. As the surgical recovery of organs can be usually accomplished only in circumstances of brain death, how the latter is understood will clearly affect who is a potential donor and when organs can be permissibly retrieved for transplantation. At the same time, an issue which is highly controversial in the context of a whole-brain definition—namely, recovery of organs from anencephalic infants—would clearly be far less controversial under the higher brain functions definition proposed by most of the authors in this volume. Since anencephalic infants present with only some portion of the brain stem, but with the rest of the brain absent, recovery of organs from these infants would be unproblematic when death is conceived as irretrievable loss of higher brain functions. Parents of anencephalics, who often seek to have their infants be organ donors, could then realize same positive benefit from an otherwise terrible situation. We should also then be able to provide help to others in need of organs by significantly increasing the available supply.

January, 1987   RICHARD M. ZANER

NOTES

[1] President's Commission: 1981, *Defining Death*, U.S. Government Printing Office, Washington, D.C.
[2] President's Commission: 1983, *Deciding to Forego Life-Sustaining Treatment*, U.S. Government Printing Office, Washington, D.C.

INTRODUCTION

It was mainly in view of two developments that, in 1968, an Ad Hoc Committee was formed at Harvard Medical School to examine the definition of death as involving "brain death" ([1], p. 337). On the one hand, "improvements in resuscitative and supportive measures have led to increased efforts to save those who are desperately injured." At times, these efforts were only partially successful, with the result that an individual could be left with circulation and respiration (at times, but not always, artifically supported), but whose brain was irreversibly damaged. On the other hand, obsolete criteria for the diagnosis of death could lead to controversy in obtaining organs for transplantation, which had already become sufficiently developed to challenge those criteria.

No physician wished to make the critical error of keeping a patient "alive" whose condition was so grievously compromised that all such care was in fact pointless, and whose organs could conceivably salvage some other patient's life. But neither did any physician wish to remove organs prematurely from a patient who was not yet "dead," and thus be culpable in medical, moral, and legal terms.

Inherent to these discussions, especially prior to the Ad Hoc Committee's Report, was the fact that neurological knowledge was becoming far more sophisticated, and technologies for sustaining respiration and circulation, as well as establishing the presence or absence of brain activity, were becoming more effective by the mid-1960s. With enhanced ability to establish central nervous system activity, with increasing incidence of successful organ transplantation, and with ever more successful resuscitation and maintenance of patients who only a few years before would have been lost, it had by the late 1960s become a matter of urgency to re-examine and even to redefine the medical diagnosis of death with greater accuracy.

The result was the Committee's recommendation to the medical community that the "brain death syndrome," or "irreversible coma," be adopted. The Committee was mainly concerned with those comatose individuals who had lost all discernible central nervous system activity, that is, those in whom such functioning was "abolished at cerebral, brain stem, and often spinal levels" ([1], p. 340). Even if respiration and circulation could be maintained in such patients, they would by this new

1

Richard M. Zaner (ed.), Death: Beyond Whole-Brain Criteria, pp. 1–14.
© 1988 *by Kluwer Academic Publishers.*

definition be considered dead. Accepting the Committee's recommendation, thus, would mean that "obsolete criteria" had been appropriately updated, and that it would then be permissible to remove viable
organs for transplantation from such patients, and thus to save other,
less grievously damaged patients.

Although no state had a statute defining death as of 1969, following
the Committee's recommendation many legislatures were discussing the
need for one. Through the 1970s, many began to adopt a statutory
definition incorporating the recommended "brain death" clause (in one
or another version, some including heart and lung death along with
brain death). By 1980, when the President's Commission for the Study
of Ethical Problems in Medicine and Biomedical and Behavioral Research began to study the issue, 24 states had adopted some form of
statutory definition (subsequently, at least 38 states have adopted one or
another version).

Aware that changes in biomedical knowledge and technology had
created a need for public recognition and sanction of "brain death," but
at the same time that in a majority of states neither legislatures nor
courts had yet responded to this increased public awareness, the Commission sought to secure agreement on a single, uniform statutory
definition. Acting together with professional associations in law, medicine, and state government, the Commission recommended a Uniform
Determination of Death Act (UDDA):

An individual who has sustained either (1) irreversible cessation of circulatory and
respiratory functions, or (2) irreversible cessation of all functions of the entire brain,
including the brain stem, is dead. A determination of death must be made in accordance
with accepted medical standards ([5], p. 2).

The Commission's recommendation, however, was adopted against
the strongly voiced opposition of a number of physicians, philosophers,
theologians, and others, who had become increasingly outspoken during
the late 1970s as more and more state legislatures began adopting some
version of a "brain death" statute. These disputes tended to be shaped
by issues made prominent by a certain kind of patient, most publically
symbolized perhaps by Karen Ann Quinlan: an individual who, from
traumatic or non-traumatic causes, was in a "persistent vegetative state"
without any trace of consciousness or personal characteristics (due to
irreversible damage to his/her higher brain functions).

Such individuals have suffered severe brain damage, either from direct injury to the brain or from cardiac arrest (stopping circulation to the brain). Since the cerebrum, and especially the cerebral cortex, is more easily injured by loss of blood flow (by not being properly oxygenated) than is the brain stem, even a very brief period without respiration or circulation (probably four minutes maximum) damages the cerebral cortex permanently. Thus, the more resistent brain stem and cerebellum may continue to function while the cerebrum and cerebral cortex are destroyed.

Such individuals often require mechanical ventilation, many of them for only a limited time, after which they begin to breathe spontaneously (since the regulatory functions of the brain stem and cerebellum continue). Nevertheless, even though they frequently exhibit a range of involuntary movements (facial grimaces, body posturing, etc.), they almost never recover consciousness. With artificial life-supports, however, including intravenous feeding, antibiotics, and others, such individuals can survive for quite lengthy periods (in a few cases, over 30 years).

The kind of patient which the Commission and other "whole-brain" advocates took as the paradigm for defining human death, however, is distinctly different from those in a "persistent vegetative state." For the Commission, only those individuals who have lost *all* brain functions, including those associated with the brain stem, are properly regarded as dead: that is, those who have lost not only the higher functions supporting consciousness but also the reflexes controlled by the brain stem (gag reflex, swallowing, the urge to breathe, etc.). Even though circulation and respiration in such individuals can be maintained by mechanical means for a short time, what has been lost is the central trait distinguishing between the living and the dead, "the one characteristic of living things which is absent in the dead," namely, "the body's capacity to organize and regulate itself" ([5], p. 32). Death in these terms is "that moment at which the body's physiological system ceases to constitute an integrated whole;" or, "when an individual's breathing and circulation lack neurologic integration, he or she is dead" ([5], p. 33). When this integration of the functioning of major organ systems is lost, what is left is no longer a functional or *organic* unity, but merely a *mechanical* complex ([3], pp. 389, 391).

As each of the authors included in this volume makes quite clear, "whole-brain" advocates are led to their view by two main considerations. (1) First, they seem concerned about the possibility of error in diagnosing death by means of the loss of "higher brain" functions. As D. N. Walton insists, "there is a possibility of indeterminacy or error" ([8], p. 50) in such diagnoses, and because of that it is always the better part of wisdom to err on the safe side (tutiorism). We know that if the whole-brain is destroyed, the higher brain functions are also destroyed; but we cannot safely tell when just the latter have been destroyed; therefore, the whole-brain formulation is preferable, because it is safer. The Commission, too, seems to have been persuaded by such a view. Thus, in his article in this volume, Alexander Capron, the Executive Director of the President's Commission, at one point argues that

> . . . however telling it may seem rhetorically to say that "if one small reflex [e.g. the gag reflex] were all that were left of an organ's functioning that could not be human life," still, as a practical matter, this seemingly odd position would have to be tolerated if the only way to diagnose the existence of more significant functions involves clinical measures that also encompass the less significant ones. . . . ([4], pp. 158–159).

That is to say: even such a patently silly conclusion seems warranted because, "as a practical matter," the irreversible cessation of higher brain functions can be determined only when the whole-brain has ceased, irreversibly, to function. *That*, at least, is the only way we can be *certain* an individual is really and truly dead.

(2) Second, then, the whole-brain advocates want to emphasize the *practical* matter of social policy (and, of course, legislative statute). Since, at least at present, "neither basic neurophysiology nor medical technique suffices to translate the 'higher brain' formulation into policy" ([5], p. 40), while the whole-brain formulation does, only the latter is acceptable. After reviewing all relevant considerations, the Commission concluded that

> it is not known which portions of the brain are responsible for cognition and consciousness; what little is known points to substantial interconnections among the brain stem, subcortical structures and the neocortex. Thus, the "higher brain" may well exist only as a metaphorical concept, not in reality. Second, even when the sites of certain aspects of consciousness can be found, their cessation often cannot be assessed with the certainty that would be required in applying a statutory definition ([5], p. 40).

Thus, since the only presumable "certainty" is the assessment (i.e., medical tests or measurements) of the loss of the "whole-brain" (which of course includes those "higher brain" functions), the only practical course for policy definition is to go with what can currently be done. More important, for the Commission and other "whole-brain" advocates, is the plain fact that the underlying premise for every "higher brain" formulation is deeply problematic. There have been two main views here. One such formulation seeks to define death as the loss of what is essential to being a person: characteristics such as thinking, reasoning, human intercourse, which distinguish human beings from other creatures, are also directly related to "higher brain" functions. When those are lost, the "person" is dead. Another formulation emphasizes the persistent *identity* of a person through moments of time, an identity which is irretrievably lost when certain "higher brain" functions are destroyed, since that identity is dependent on their continuation.

For something as significant culturally as a uniform statutory definition of human death, one requirement must be that there is a general consensus over what constitutes either "personhood" or "personal identity." The fact is, however, the Commission believes, there simply *is no* such consensus. Neither philosophers, physicians, nor the general public shows any substantial agreement about what essentially constitutes a "person." And, "despite the best attempts by philosophers to solve it," the problem of personal identity "has persisted for centuries" and continues to be disputed today ([5], pp. 39–40). To press for a "higher brain" formulation, then, would be to advocate a "wholly new concept of death," a concept which could only be based on quite arbitrary considerations, given the failure to achieve any significant societal consensus.

This line of argument, defended by Capron here, is also directly attacked by Roland Puccetti, Edward Bartlett and Stuart Youngner, and Robert Veatch in their articles, while Marx Wartofsky and Patricia White and their commentators raise a number of legal, social, and other issues associated with the dispute. In the lead article of this volume, Martin Pernick provides the richly detailed historical context required for a proper understanding of all these issues.

The main arguments against the whole-brain formulation may be briefly summarized. The tutioristic ("safe side") argument, such as it is, may be quickly disposed of: after all, if we really wished "to be safe" in determining when a person has died, only total somatic death would be acceptable, for only then would there be anything like "certainty." At the same time, of course, we should have undercut any significant prospects of organ transplantation, since complete somatic death would have to include total cessation of every organ and every cell. That requirement, Puccetti argues, is not only silly; the entire tutioristic point is quite irrelevant ([6], pp. 76–80). Strictly brain stem reflexes (gag, swallowing, pupillary, etc.) remain intact in individuals who are in a persistent vegetative state (those exhibiting apallic syndrome), but they clearly do not alter their functioning when the higher brain processes are destroyed. Hence, they do not suggest, indicate, or imply in any way that the apallic syndrome individual continues to "feel" or "sense" anything. Thus, to be apprehensive (tutiorism) that possible "feeling" or "sensation", persists in an apallic syndrome patient is to confuse essentially different matters. The mere continued neural firing of tha-lamic pain centers is not equivalent to experiencing pain: "it is *we* who get the pains, not those structures," Puccetti correctly observes ([6], p. 80). Thus, if an individual is irreversibly comatose as a result of the destruction of the higher brain functions, the basis for a conscious and hence a personal life in this world is gone forever ([6], p. 87), even though it may well be possible to maintain the rest of that individual's body by mechanical means, and even though his brain stem continues to function (by spontaneous respiration, for instance). "Permanent uncon-sciousness is permanent unconsciousness whether the condition is asso-ciated with a body that lives by virtue of being able to breathe spontaneously or not" ([6], p. 83).

In the end, the Commission and other whole-brain advocates have opted for a philosophical view which identifies "human life" with strictly *organic*, bodily conditions: the "moment at which the body's physiologi-cal system ceases to constitute an integrated whole," in the Commis-sion's words. Bartlett and Youngner insist that the whole-brain position really results from a serious confusion. Any formulation must include (a) a concept or *definition* of what it means to die, but also (b) opera-tional *criteria* by which to determine that death has occurred, and (c) specific medical *tests* to measure whether the criteria have been fulfilled ([2], p. 200). While definitions are basically *conceptual*, criteria are primarily *physiological* standards for determining whether death, as defined conceptually, has occurred, and the medical tests function specifically to demonstrate *when* the criteria have been fulfilled.

Thus, the whole-brain advocate's appeal to the lack of current tests or assessments as a reason for rejecting the higher brain concept of death is quite mistaken: "a definition cannot be considered wrong simply be-cause we lack specific, reliable tests to demonstrate the fulfillment of its criteria" ([2], p. 200). After carefully reviewing the recent history of the whole-brain position, they note that not until quite recently did anything like a clear definition (i.e., concept) of death actually appear. This concept, however, gives almost exclusive focus to the integration of essentially reflex, vegetative functions of the brain. Or, in Puccetti's terms, it is the mainly "janitorial" functions of the central nervous system, those requiring no conscious direction whatsoever, which are given priority ([6], p. 84). Hence, for the whole-brain advocates, it is the biological organism (or, more specifically, the physiological/anatomical nervous system) which is definitive for life and for death, not the *person* whose organism (or nervous system) it is. In effect, this position has thus *evaded* the central issue (*who*, after all, had died?); or, perhaps more accurately, it has tried to develop a concept (definition) of death from a list of *criteria* and medical *tests* – which clearly puts the cart before the horse.

What is needed, to the contrary, is a sense of the *qualitatively* distinct functions of the human central nervous system, for not all such functions are equally significant for human or personal life. Instead, through its quite frank reductivism, it has effectively muted, if not obliterated, such qualitative distinctions in favor of merely *quantitative* ones. To establish which brain functions are more important than others, however, "we need a standard of selection" ([2], p. 209), and just this is lacking in the whole-brain formulation. The latter invariably winds up "describing the death of something that is not a human being," Bartlett and Youngner insist ([2], p. 209) that is a merely biological entity bereft of any distinctly human, personal presence. While it is surely true that such entities can and do die, it is equally and more importantly true that persons also die. Indeed, the permanent loss of personhood *is* death in the most significant sense, and it is this that should form the core of public policy or legislative statute.

Like others included in this volume, Robert Veatch has long been opposed to the whole-brain concept of human death. In the article included here, he moves the argument to somewhat different grounds.

In the first place, he argues, what passes for arguments by the Commission are in fact little more than non sequiturs. Thus, the fact that there may be a lack of consensus about the meaning of "personhood" or "personal identity" is surely no argument against a higher brain concept of death. If it were, not even the whole-brain formulation would be acceptable, since there continues to be wide dispute over what constitutes human death ([7], p. 175). On the other hand, as both Puccetti and Bartlett and Youngner emphasize, there are very good conceptual and philosophical reasons to reject the whole-brain concept, the same reasons which suggest that the higher brain concept is much sounder.

It is also a non sequitur to be worried that persons with apallic syndrome "could be buried or otherwise treated as dead persons," as the Commission stated ([5], p. 40). If the "person" has died, after all, then there are hardly good reasons to continue treating him as if he were still alive! Most of us, obviously, would wish to postpone such actions as burial until certain residual functions (heart-beat, breathing, etc.) have ceased, but this is hardly an argument against considering such persons as dead.

While the tutioristic concern might be relevant to the problems of policy formulation, this is not, as both Veatch ([7], p. 177) and Puccetti ([6], p. 85) point out, an argument against a higher brain concept of death. Finally, the Commission's expressed concern that the latter concept is really a "radically new definition" of death while the whole-brain formulation is not, is simply mistaken: why turn to the "brain" at all if not because those neurological functions specifically support *personal* life? That is, if there really is a new development, it occurred with the shift from the heart/lung formulation to the brain, and not, as the Commission has claimed, in that proposed by the higher brain function advocates.

Veatch's own view continues to be that when the "embodied capacity for consciousness or social interaction is gone irreversibly, then, and only then, do I want society to treat me as dead" ([7], p. 182). Acknowledging that this is only one of many possible religious/philosophical views, and insisting that his view should not be imposed on other people (any more than any other), Veatch concludes that individuals in our pluralistic society should be left free to examine various views for themselves, adopt the positions that are most plausible to them, and thus be free to choose a definition of death commensurate with such a view.

Patricia White, on the other hand, argues that it may be misleading to place so much weight on determining whether someone is alive or dead, that is, simply "to *assume* that there is a specifiable point of division between life and death and that all hard cases fall on one side or the other" ([9], p. 104). Rather, she suggests, it may well be that under any relatively exact statutory definition of death the hard cases remain quite *indeterminate*, despite that effort to be legally exact. In fact, she **argues**, the cases which force us to think hard about whether an individual is dead or alive are most often those where the person himself has comparatively little at stake, "for they are instances where the person has permanently lost all cognitive life" ([9], p. 103). Since, however, these same cases do have consequences for *other* people – family, friends, physicians, etc. – and these consequences are likely to be considerably important, the really serious and difficult question is whether a person with apallic syndrome is to be treated *as if* he were alive or dead.

The difficulty here is this: greater legal and statutory exactness in defining that "specifiable point of division between life and death" effectively *mandates* that such hard cases must fall either within or outside the boundaries of definition (the person is *either* alive *or* dead, and this cannot be indeterminate). But, at the same time, such types of definition, by ruling out actually indeterminate cases (or by making them appear to be determinate – as falling on one or the other side of definitional boundary) force the law into issues that are essentially *moral*. An almost brain dead, but clearly irreversible and incurable patient, one who falls on neither side of that boundary, forces the question whether he can be treated *as if* he were dead (thus, whether life-supports can be removed). These are, however, moral issues, even while they are made to appear as legal ones. These issues, and especially the legal problems that need address because of uncertainties and inconsistencies in current law, are exhaustively explored by David Smith. Proposing legal acceptance of a neocortical definition of death as far more coherent and humane, Smith discusses the legal issues of his proposal in a thorough and fascinating way.

While there are, of course, a number of important issues occasioned by and underlying these disputes, one has clearly special significance. This concerns the "person" in relation to any definition of death.

Whole-brain formulations are deliberately conceived as basically *physiological*: death is the irreversible cessation of the physiological integration of the organism. Despite this deliberate restriction of the definition of death to the organism, however, the notion of "person" apparently cannot be excluded. Thus, the Commission defines death as "that moment at which the body's physiological system ceases to constitute an integrated whole" ([5], p. 33). Yet, in the very next paragraph, it is stated that "a *person* is considered dead . . . even if oxygenation and metabolism persist in some cells or organs" (ibid. [emphasis added]). Acknowledging that the brain is the "sponsor of consciousness" and the

integrator of bodily functions, however, the *Report* then asserts that
death concerns the "organism's responsiveness to its internal and external environment" (p. 36) – and not, apparently, the *person's* responsiveness. That view is again restated as the "collapse of the organism as a
whole," a collapse that can be diagnosed through loss of brain functions
as well as loss of cardio-pulmonary functions (p. 37).

The higher brain formulations are then rejected, primarily for a lack
of consensus on the meaning either of "personhood" or "personal
identity". In its later consideration of the four levels of generality at
which public policy can be stated, the Commission expressed its view
that only the level of "physiological standards" (p. 56) is appropriate for
that purpose. This whole discussion is, however, seriously confused, and
it is again the place of the "person" that is central.

The first level of generality is conceptual or philosophical. Several
examples are offered, among them the "permanent cessation of the
integrated functioning of the organism as a whole," and "irreversible
loss of personhood." The former is obviously the Commission's own
preferred definition, while the latter is one version offered by higher
brain definitions. Yet, *both* of these, and any other "basic concept" of
death is here said to be of "little concrete help in the practical task of
determining whether a *person* has died . . ." (pp. 55–56 [emphasis
added]). With this, the confusions multiply.

(1) Helpful practically or not, the issue is said to be the "death of the
person" – contrary to the Commission's own deliberate exclusion of
"person." (2) In any event, the Commission excludes not merely a
"personhood" concept, but also *its own physiological* concept, as being
of no concrete help in determining whether a person has died – again,
contrary to its own earlier argument. Such basically conceptual discussions, it is asserted, "would lead down arcane philosophical paths which
are at best somewhat removed from practical application in the formation of law" (p. 56). So, *both* concepts are little more than "arcane" and
impractical.

(3) It is also stated, however, that when physicians used to apply the
more traditional heart and lung functional standards, "they were
affirming only that the loss of those functions *indicated* that a person had
died" (ibid.). Death is the death of the person, not the body. It is then
asserted that new technologies which interfere with these indicators
(such as the artificial respirator) "do not necessitate a change in concepts, provided that alternative indicators of the current concept are

available," and the brain-oriented indicators discussed earlier in the *Report* are said to provide such alternatives (pp. 56–57). "Indicators" or not, however, it is again stated that it is the death of the *"person"* that is at issue.

(4) Indeed, that the "person" is in fact the issue is expressly affirmed on the next page of this Chapter Five, in a section curiously entitled *"Death of the Organism as a Whole"*:

The death of a human being – not the "death" of cells, tissues or organs – is the matter at issue. The cessation of vital bodily systems provides the basis for broad standards by which death can be judged to have occurred. But such functional cessation is not of interest in and for itself, but for what it reveals about the status of the person. What was formerly a person is now a dead body and can be socially and legally treated as such (p. 58).

As will be seen in the essays that follow, not only the Commission but each of the major whole-brain advocates is taken to task either for inappropriately ignoring the "person," or for inconsistently including it, or for advocating a basically reductivist position which renders human death incoherent. What can be seen from the above passages of the Commission's *Report*, moreover, is that its own deliberate effort to exclude the "person" from the definition of death by restricting death to the physiologic integrational functions of the brain does not succeed, since the "person" is in fact brought back in as the "matter at issue". After all, the Commission expressly insists, brain functions (*whatever* these may be) are significant, *not* in and of themselves, but solely for what they reveal about the "status of the person".

Now, in these terms, several issues assume central import. First, Bartlett and Youngner emphasize, whole-brain enthusiasts have in fact confused two basically different levels: that of the *concept* of death and that of the *criteria* by which it is to be determined that death has occurred. Thus, when the Commission states that the loss either of heart/lung or of brain functions is an indicator (i.e., "criteria") that a "person has died," this seems consistent only if their basic argument is accepted. Indeed, the Commission's emphasis that functional cessation is significant "for what it reveals about the status of the person" (i.e., that what *was* a "person is now a dead body") not only confirms Bartlett and Youngner's very point, but also stands in clear contradiction to the Commission's own arguments.

Second, precisely because the loss of either heart/lung or brain functions is an indicator that a "person has died," it is conceptually *incoherent* to argue that death is merely the moment at which the body's physiological system ceases to constitute an integrated organic whole. Rather, by the Commission's own admission that the "status of the person" is the key issue for any definition of death, the "integration" mentioned can only be that which otherwise obtains between the "person" and his/her embodiment by a specific organism. As Veatch suggests, it is the "embodied capacity for consciousness or social interaction" that is central. What occurs at death, in short, is the cessation of this embodiment, i.e., the irreversible cessation of the integration of the embodied person "as a whole" and not merely the "organism as a whole."

A third issue then becomes prominent. The Commission had expressed great concern that there must be a "broad social and professional consensus" on the "basic concept" of death, and had indicted higher brain formulations for failing to have that consensus, especially for the purposes of social policy and legislative statute.

On a matter so fundamental to a society's sense of itself – touching deeply held personal and religious beliefs – and so final for the individuals involved, one would desire much greater consensus than now exists before taking the major step of radically revising the concept of death ([5], p. 41).

To this it can be answered, (a) that if brain functions are "indicators," so are the proposed higher brain functions *indicators* of personhood – and their loss the "indication" that "a person has died," in the Commission's own language. If (b) "the status of the person" is in truth what is significant in assessing the meaning of the loss of such brain functions, then the important issue is the detemination of those brain functions that are specifically correlated with (and thus "indicators" of) those characteristics that constitute the "person". That there may not at the present time be adequate or accurate tests by which to measure the loss of those specific functions is of course an important issue, but it does *not* concern the "basic concept" of death. (c) The idea that there are specific functions of the brain which support, are correlated, or associated with specifically "personal" characteristics, is thus not at all a "radical" idea. Therefore, (d) since the Commission itself clearly affirms the significance of the "person" for any definition of death, it faces the issue of social consensus quite as much as any other effort to define death.

It is unfortunate that the Commission's *Report* not only confused very different issues but wound up endorsing a reductivist, physiological rationale for its definition of death that is clearly inconsistent with what is acknowledged to be "the matter at issue," namely, "the status of the person". It may well be, as Veatch urges in his article ([7], pp. 183–185), that the special purposes of social policy require, *at present*, that a "whole-brain" formulation be utilized. It may, on the other hand, be true, as Puccetti ([6], pp. 85–86) argues, that we already have quite adequate and accurate means by which to assess the loss of those higher brain functions that support consciousness, and therefore that statutory definitions ought to recognize this formally. In any event, it is inarguable that death in the only significant sense for human beings is the death of the person: "What was formerly a person is now a dead body and can be socially and legally treated as such," in the Commission's own terms ([5], p. 58.)

At the very least, then, the essays that follow clearly succeed in their collective insistence that the central issues inherent to any definition of death must be kept rigorously open – especially that concerning the necessary medical and biomedical task of determining those portions of the brain which are correlated with consciousness and the "person" (along with that of developing specific tests for measuring when the criteria or "indicators" have been fulfilled). As Bartlett and Youngner emphasize, furthermore, the key to the entire discussion is the *conceptual* or *philosophical* issue: to be capable of developing criteria ("indicators"), much less specific assessments ("tests"), one must have "a standard of selection" already at hand ([2], p. 209).

Clearly, much remains to be done physiologically regarding the correlation between portions of the brain and consciousness, as well as technically to develop tests by which to assess their irreversible loss. At the same time, much remains to be done to clarify the conceptual issues, in probing and clarifying what that "standard of selection" must be. A significant part of this task must be specifically devoted to an issue that has unfortunately remained quite confused in most of the literature on brain death.

We have noted that those inclined toward a "whole-brain" formulation have nonetheless made regular reference to the "person" and to "consciousness" (cognition, affection, etc.). Those persuaded by a higher brain definition, on the other hand, not only use these terms, but others as well: "Personal identity," "self," "human being" to mention but several. It must be asked: do these various terms refer to the same or to different things? If one reads the literature on "consciousness," "persons," or "human being," on the other hand, one finds an even greater array of such terms: "agent," "subject," "subjectivity," "ego," "soul," and "spirit," to mention only a few (*see* [10], pp. 92–114, 144–173 for a discussion of these). These broader, philosophical issues, are admirably explored by Wartofsky, Lachs, and their commentators.

Clearly, a central part of the conceptual issue focused on developing a sound "standard of selection" must grapple with this issue directly. The present volume, in any case, succeeds admirably not merely in keeping the issues open, but in fact succeeds in advancing the discussion in important ways. A careful study of these essays must give the serious reader pause for reflection about the need for a coherent concept of human death, as it must make us reconsider the content and purposes of public policy and statutory definition, concerned not only with human death, but organ donation and transplantation (in particular, those involving anencephalic infants).

RICHARD M. ZANER

Vanderbilt University
Nashville, Tennessee

BIBLIOGRAPHY

1. Ad Hoc Committee of the Harvard Medical School to Examine the Definition of Death: 1968, 'A Definition of Irreversible Coma', *Journal of the American Medical Association* **205**, 337–340.
2. Bartlett, E. T. and Youngner, S. J.: 1988, 'Does Anyone Survive Neocortical Death?', this volume, pp. 199–215.
3. Bernat, J. L., Culver, C. M., and Gert, G.: 1981, 'On the Definition and Criterion of Death', *Annals of Internal Medicine* **94**, 389–391.
4. Capron, A. M.: 1988, 'The Report of the President's Commission on the Uniform Determination of Death Act', this volume, pp. 147–169.
5. President's Commission for the Study of Ethical Problems in Medicine and Biomedical and Behavioral Research: 1981, *Defining Death: Medical, Legal, and Ethical Issues in the Determination of Death*, U.S. Government Printing Office, Washington, D.C.
6. Puccetti, R.: 1988, 'Neocortical Definitions of Death and Philosophical Concepts of Persons', this volume, pp. 75–90.
7. Veatch, R. M.: 1988, 'Whole-Brain and Higher Brain Related Concepts of Death', this volume, pp. 171–186.
8. Walton, D. N.: 1981, *Brain Death: Ethical Considerations*, Purdue University Press, West Lafayette, Indiana.
9. White, P.: 1988, 'Should the Law Define Death?, this volume, pp. 101–109.
10. Zaner, R. M.: 1981, *The Context of Self*, Ohio University Press, Athens, Ohio.

PART I

HISTORICAL AND CONCEPTUAL FOUNDATIONS

MARTIN S. PERNICK

BACK FROM THE GRAVE: RECURRING CONTROVERSIES OVER DEFINING AND DIAGNOSING DEATH IN HISTORY

On December 2, 1967, a car accident in Cape Town, South Africa, destroyed the brain of a young woman named Denise Ann Darvall. Dr. Christiaan Barnard and his colleagues implanted her heart in Louis Washkansky, a victim of terminal heart failure, thus making him the world's first successful heart transplant recipient. In reporting the dramatic event, both the professional and popular press immediately focussed on the question asked by the headline in *Newsweek Magazine*'s coverage of the story: "When Are You Really Dead?" ([245], [242], [243], [184], p. 59).

Most of these accounts made defining death seem as unprecedented as heart transplantation itself – a radically new dilemma created by new medical technology. This paper will argue that such common assumptions are wrong on at least three crucial points. (1) The modern difficulty in deciding who is dead is only the most recent resurrection of an issue that has revived repeatedly throughout medical history. (2) In the past, as now, difficulties in defining and diagnosing death often did spring from new medical discoveries, especially in such areas as experimental physiology, resuscitation, and suspended animation. The intimate connection between such discoveries and past clinical doubts about death has not previously received adequate recognition. But, (3) the resulting debates, past and present, were the product of a complex interplay of social, professional, and ethical changes, and cannot be understood as simply the result of new medical knowledge. Death has never been completely definable in objective technical terms. It has always been at least in part a subjective and value-based construct.

I. DEATH FROM ANTIQUITY TO THE SIXTEENTH CENTURY: A BRIEF OVERVIEW

Once upon a time, long before modern machinery, everyone agreed that death occurred when your heartbeat and breathing stopped. This appealing fable is true to a point, but it leaves unanswered many vital questions. Were the heartbeat and breath defined as constituting life

17

Richard M. Zaner (ed.), Death: Beyond Whole-Brain Criteria, pp. 17–74.
© 1988 *by Kluwer Academic Publishers.*

itself, or were they merely physiological indicators that life was present? Which of the two was more important, and were there any other vital events? What tests, if any, could determine whether these functions had actually ceased? What was the connection between the death of an organism-as-a-whole, and the death of its discrete body parts? What was the relation between the end of physical vitality and the end of personal existence? And for what purposes did people seek the answers? From antiquity to the present, all these issues have remained extremely controversial.

The combined observations of surgeons, warriors, butchers, and executioners led most ancient societies to conclude that an organism's body parts did not always die simultaneously, and that the vitality of certain specific organs was crucial for the life of the whole. Classical Greek physicians believed that death could begin in the lungs, brain or heart, but that the heart alone served as the seat of life; the first organ to live and the last to die. The heart created the vital spirits, the essence of life. Thus, the heartbeat alone distinguished between the living and the dead. Breathing merely regulated the heat of the heart ([59], Vol. 1, pp. 229–236; [2], p. 19; [251–252]).

The brain did have a vital role. Hippocrates located reason, sensation, and motion in the brain; a view shared by the anatomists of Alexandria and by Galen. The early Church Fathers even localized specific mental faculties in specific ventricles of the brain ([60], pp. 296–298). But the heartbeat remained the sole indicator of life and death.

In the Hebrew tradition, *ruach* or breath was primary, often defined as constituting life itself. This view remained a strong influence in Christian thought as well, at least through the Middle Ages (e.g., Psalms 104:29; [199], pp. 280, 285; [10], p. 248). Even within Judaism, however, the role of breath was not completely unchallenged. By the fourth century, a minority of Talmudic sages accepted the heartbeat as a valid alternative indicator of life and death ([199], p. 299). And Maimonides, a twelfth-century rabbi and physician steeped in Greco-Arab medicine, asserted the vital significance of the head. He considered a decapitated body dead, even if it still moved, because its motions lacked the central direction that presumably indicated the guidance of a soul ([199], pp. 297–298). Thus, various classical and medieval traditions recognized certain organs as crucial to the life of the organism-as-a-whole, yet there was much disagreement over which specific organs and functions were vital, and why.

Furthermore, "death" meant not only the loss of bodily vitality, but the end of personal existence as well. The precise relationship between these two kinds of death proved especially controversial. Does the life of the self depend on the life of the body, or can the personality survive, or predecease, the organism? Is personal existence a unitary, all-or-nothing phenomenon? Or is it, like the body, composed of distinct components whose existence might be terminated separately?

Plato taught that both the vitality of the body and the identity of the individual depend on the presence of an immortal soul, distinct in nature from the body. A person is his or her soul; the body is simply a tool or vessel used by the soul ([96], Vol. 1, p. 98; [77], pp. 34–71; [218], pp. 5, 15). Aristotle, however, held that a person is not simply a soul, but an integral combination of soul and body. The soul is to the body as the impression of a seal is to wax (or as the smile to the Cheshire cat) ([218], pp. 4, 6, 10). The soul cannot exist without the body (except perhaps for that part of the soul which is pure intellect) ([96], Vol. 1, p. 110; [77], pp. 72–81). The death of the body is therefore the death of the soul.

But Aristotle sharply distinguished those parts of the soul responsible for bodily vitality (the vegetative soul of nourishment and growth) from those governing motion and sensation (animative soul), and the mind (rational soul) ([96], Vol. 1, p. 108; [77], p. 77). While the proper functioning of the vegetative soul was essential for bodily life, rationality conceivably might terminate without causing complete bodily death; indeed, animals and vegetables lived their entire lives without rational souls. In short, for Plato, the unitary, holistic self always survived the death of the body. For Aristotle, a living person required a living body, but life consisted of three discrete and hierarchical functions, each governed by a separate level of the self.

Different classical and medieval physicians not only differed in their definitions of life, and their views on the physiological indicators that signified life had ended; they also recognized cases in which the attempt to apply their definitions and indicators resulted in diagnostic errors. "[S]o uncertain is men's judgment," wrote Pliny, "that they cannot determine even death itself." Galen listed hysteria, asphyxia, coma, and catalepsy among the conditions he thought could suspend all signs of life for weeks without precluding recovery ([222], pp. 492, 496; [254], pp. 10, 104, 118–121). St. Augustine knew a monk, aptly named Restitutus, who could suspend his own heartbeat ([12], Bk. XIV, Ch. 24; IV, p. 391). Despite the centrality of breath in the Jewish tradition, Maimo-

nides and Rashi both considered it possible to survive protracted
drowning. The Bible credited Elijah with restoring breath to a corpse
([199], pp. 286, 305). Ancient Egyptians and Romans produced such
volatile caustics as sal ammoniac and spirit of hartshorn, capable of
restoring the vital functions in cases of syncope ([94], p. 174; [222], pp.
496, 499). Thus, even the absence of heartbeats and breathing did not
always mean death.

However, we must remember that diagnosing death was not usually a
clinical responsibility for the classical doctor, due to Hippocratic medical ethics and its prohibition of medical treatment for terminal patients.
The doctor's duty was to forecast an impending demise and then
withdraw from the case, not to remain in attendance long enough to
diagnose or certify actual death ([167], p. 105). Thus, when classical and
medieval physicians spoke of the *proprietates mortis*, the "signs of
death," they did not mention heartbeats, pulse, or breath, but repeated
the portrait of impending death painted in Hippocrates' *Prognostikon*
[195]. Then as now, the "signs of death" indicated when the doctor's job
was finished, but for classical physicians these "signs" did not mean the
patient was actually dead. For most of Western history, the actual
diagnosis of death was primarily a non-medical function.

Various classical texts indicate practices employed to protect against
erroneous diagnoses. The Talmud, for example, records that ancient
Jewish custom was to visit a corpse in its crypt for three days after death
to check for renewed vital signs ([199], p. 285). During major epidemics, like the Black Death of 1348 and succeeding plague outbreaks,
when such careful precautions were unsafe and impractical, the popular
dread of being buried alive rivalled the terror of the disease itself ([254],
p. 50). Thus, from the very beginning of Western history, defining and
diagnosing death proved both perplexing and controversial.

II. 1740–1850: MEDICAL TECHNOLOGY, SOCIAL VALUES, AND THE PANIC OVER PREMATURE BURIAL

Death was never easy to define or diagnose, yet some historical periods
demonstrated far more discomfort over these uncertainties than did
others. One era of particularly intense concern began about 1740 and
lasted through the middle of the next century or longer, as the scientific,
social, and ethical effects of the Enlightenment combined to render the
boundary between life and death frighteningly indistinct.

In the mid-seventeenth century, the Papal physician Paulus Zacchias had written that no sign prior to the start of putrefaction could reliably distinguish the dead from the living ([126], p. 42; [88], p. 272; [227], p. 409). But not until the following century did such comments evoke much concern. In 1740, the eminent Franco-Danish anatomist Jacques Bénigne Winslow published a Latin dissertation on *The Uncertainty of the Signs of Death and the Danger of Precipitate Interments and Dissections* [253]. A French translation with extensive additions by Jacques-Jean Bruhier d'Ablaincourt followed in 1742. Two English editions of Bruhier's version, with further additions by M. Cooper, appeared in 1746 and 1751. A second French edition was expanded to two volumes. Within two decades, Italian, Swedish, German and other translations had proclaimed the uncertainty of death across all of Europe.[1] Although one noted surgeon, Antoine Louis, asserted that the onset of rigor mortis was a sufficient diagnostic sign, Diderot's *Encyclopédie* endorsed Winslow and Bruhier's assertion that only putrefaction provided certain proof of death ([3], p. 26). More than fifty medical works on the subject appeared before 1800; by 1850 the books and articles could be counted in the hundreds ([227], pp. 409–436). Anglo-American Robley Dunglison's pioneering medical dictionary of 1833 agreed that "the only certain sign of real death is the commencement of putrefaction" ([68], Vol. 2, p. 49),[2] while others doubted that even incipient decomposition could be distinguished reliably from such diseases as gangrene.[3]

The most dramatic result of physicians' lack of confidence in diagnosing death was the public panic over "premature burials." Today, this phenomenon is best remembered through such literary creations as Edgar Allen Poe's *Fall of the House of Usher*, and "The Premature Burial" [174], [122]. In fact, tales of living entombment can be found in French literature by the seventeenth century ([10], pp. 376–378); early nineteenth-century magazines were filled with models of the genre ([176], Vol. 1, pp. 177, 338; [122]).[4]

But burial alive was not simply a literary device. Public concern was real and intense. Both individuals and governments reacted with growing alarm.

Resuscitation, Suspended Animation, and the Definition of Death

Why did eighteenth-century society suddenly come to doubt its ability to distinguish the quick from the dead? A series of startling medical discoveries played a little-known but central role, by revealing that many people diagnosed as dead could in fact be revived.

The most important of these new discoveries was artificial respiration.

Although previous historical studies have overlooked the significance of the connection, knowledge of artificial respiration and uncertainty over the signs of death spread virtually simultaneously. The pioneering medical works on each subject both appeared in Paris in 1740 [186, 253], and, every subsequent edition of Winslow's treatise on the uncertainty of death added extensive new selections from the latest works on resuscitation.[5]

Beginning in 1767, physicians and reformers across Europe and the Americas organized what they called "humane societies" to teach artificial respiration, resuscitate victims of drowning and suffocation, and promote research on new life-restoring techniques.[6] The London Society alone claimed to have revived over 2,000 people by 1796 ([3], p. 31 n.19). William Hawes, the medical practitioner who founded the London Society in 1774, shortly thereafter wrote one of Britain's earliest works on premature burial, in which he used the humane society's resuscitation case records to prove that nothing short of putrefaction could distinguish death from life ([103], p. 40; [128], p. 422; [61]).

Even physicians who ignored the specific fear of premature burial began to doubt their ability to diagnose death after learning about artificial respiration.[7] By 1792, one writer could marvel, "The time is within the recollection of many now living, when it was almost universally believed, that life quitted the body in a very few minutes after the person had ceased to breathe" ([164], p. 5). The discovery of artificial respiration thus played a key role in undermining faith in the ability to diagnose death accurately.

Furthermore, eighteenth-century resuscitation also included measures to revive circulation, sensation, and motion, as well as respiration. Drowning victims were subjected to vigorous shakes and thumps explicitly intended to restore motion to the blood [232, 79], ([254], p. 82). Smelling salts, to restore the vital signs in cases of syncope, first entered the pharmacopia in 1721, following their mid-seventeenth century discovery by Sylvius of Leyden ([94], p. 174).

But electric shock, not chemical or manual stimulation, dominated research on restoring heart, nerve, and muscle functions. Giovanni Bianchi used electricity to resuscitate a dog in 1755; the first human case followed in 1774. Contributors to the field included such noted scientists as Daniel Bernouilli, Alexander von Humboldt, and John Hunter. The leading exponent of electrical cardiac resuscitation was Giovanni Aldini, professor of physics at Bologna and nephew of Luigi Galvani.[8]

Electroresuscitation came too late to influence Winslow and Bruhier in the 1740s, but by the early 1800s it had joined artificial respiration in convincing physicians to question the signs of death ([159], p. 192). Aldini's public exhibitions spread such uncertainties beyond the medical community as well. Londoners were astonished by the twitching and wheezing Aldini evoked from the electrified corpse of an executed convict in 1803 ([203], pp. 361-362). Mary Shelley explicitly mentioned Aldini as a source for her 1818 account of a similar physician, named Frankenstein. Her work vividly dramatized both the ethical issues and the emotional terrors aroused by new resuscitation techniques, and the uncertainties they evoked about the distinctions between life and death ([10], p. 389).

As those distinctions became fuzzier, physicians increased their efforts to understand the long-puzzling phenomenon of suspended animation, and such related conditions as coma, catalepsy, asphyxia, and syncope. By the mid-nineteenth century, researchers identified a long list of traumas and diseases that seemed capable of counterfeiting death: extreme cold, opiates, alcohol, hemorrhage, apoplexy (stroke), suffocation, high fever, and head injury.[9] The introduction of inhalation anesthesia in 1846 added yet another new condition to the compilation [222], ([167], Chapter 3).

Research on death-like states in lower animals provided even more troubling examples. As early as 1701 Anton von Leeuwenhoek had reported the ability of many single-celled organisms to revive after months of dessication and seeming lifelessness. Baker and Spallanzani dramatically extended such findings by reviving worms after several decades of suspended animation. Claude Bernard described similar "latent life" in plant seeds. Others proved the ability of reptiles and fish to survive long periods of complete freezing ([18], pp. 537–539). The scientific study of hibernation also began in this period, with the work of John Hunter and Marshall Hall, both of whom made contributions to artificial respiration as well ([222], p. 503; [225], p. 804; [88], pp. 344–348).

More controversial still were those states of suspended animation deemed psychological or idiopathic: catalepsy, ecstasy, and trance. In Paris in the 1780s, Franz Anton Mesmer induced in his followers a trance state he attributed to "animal magnetism." Mesmer believed such trances enabled the soul to leave the body, during which time the body might either remain apparently lifeless, or move about in a zombie-like state without consciousness or will. Mesmerism not only produced apparent death; by 1784 it was also claimed to have restored the dead to life ([51], p. 58; [25], p. 57).

This alarming growth in the number of death-like states seemed to require one of two possible responses. The fact that apparently unconscious, breathless, and pulseless bodies could be revived might mean that these physiological functions were not really essential to life. "[I]t is well known," asserted one doctor in 1835, that "the circulation . . . is not essentially necessary to the preservation of the vital spark" ([209], pp. 214–216). If so, these functions could no longer be used as the criteria of human death; new indicators, and perhaps a new definition of death would be needed. On the other hand, maybe the vital functions never fully stopped during such cases. Perhaps they really continued but at a level below the sensitivity threshhold of the available means of detecting them. In that case, all that was needed was better diagnostic testing to distinguish between minimal vital functioning and death. No change in indicators or definitions would be necessary.

Eighteenth-century writers who adopted this latter view, and denied the reality of true suspended animation, included Winslow, and Giovanni Lancisi, physician to Pope Innocent XI ([254], p. 12; [126], pp. 42–44). Their view gained wide medical acceptance in the 1830s and 1840s. As Robley Dunglison explained, "When the vital functions cease, life is extinct" ([67], p. 4).[10] The problem was simply to devise more sensitive and reliable tests to detect these functions. The search for such procedures filled the literature. Tests for respiration included such traditional methods as holding a mirror, candle, or feather to the nose;[11] submerging the body and watching for bubbles ([215], pp. 138–146); and putting a bowl of liquid on the chest ([205], p. 529); as well as such new techniques as auscultation with the recently invented stethoscope ([78], Vol. 1, p. 380); and holding an hygrometer to the nose ([215], pp. 138–146).

Tests for circulation ranged from feeling the pulse manually, or listening with the stethoscope,[12] to opening an artery ([215], pp. 138–146). Physiological signs of circulatory failure included: livid spots ([2], p. 21); pallid skin ([213], p. 12); depressed loins; and sunken eyeballs ([209], p. 217; [225], p. 804, 807). A slack jaw, unblinking eye, and relaxed sphincters traditionally indicated loss of muscular irritability. Coldness signified the loss of "vital heat" ([213], p. 12; [2], p. 21; [209], pp. 216–218). Early nineteenth-century physicians used chemicals and heat to produce skin blisters, to vent the body's excesss "irrita-bility" in disease; thus the failure of irritating stimuli to produce an inflammatory response meant the vital irritability of the body had been exhausted ([40], pp. 105–106).

Failure to respond to artificial respiration soon gained acceptance as a (partly tautological) criterion of death. "Diligent use of restorative means . . . will be the best and surest mode of discriminating between Asphyxia and death," wrote one forensic medical authority ([212], p. 19). Likewise, failure to twitch in response to electrical resuscitation seemed useful as a test for loss of neuromuscular function. This distant ancestor of the flat brainwave test was first suggested by Charles Kite of the London Humane Society, in 1788 ([124], p. 125).[13] Unsuccessful use of other revival techniques, from smelling salts ([215], pp. 138–146) to blowing a trumpet in the ears ([2], p. 21), also figured in the list of death tests suggested.

Each of these and many more had their advocates, but only rigor mortis and putrefaction won any general acceptance, and even these were widely challenged. The response was to search for even more signs and tests. In 1837, Pietro Manni, toxicologist at the Universities of Naples and Rome, donated 1,500 francs to the Paris Academy of Sciences, to be awarded for the discovery of a definitive death test. The honor was bestowed on Eugene Bouchut in 1846, for the claim that two to three minutes of careful auscultation with the stethoscope unfailingly tested for the absence of cardiac function, and that heart motion was a certain indicator of life or death. This prize was only the first of more than a dozen such awards, from three separate bequests, given to the promoters of various death tests, though the majority of these prizes were not awarded until later in the nineteenth century ([137], p. 18; [203], p. 370; [3], p. 30).

However, if one or more vital functions could truly be suspended, adding more or better tests would not suffice. Those who believed that circulation and respiration completely stopped during suspended animation included Zacchias and Bruhier ([126], pp. 42–44; [254], pp. 98, 128–130), along with most of the eighteenth-century British pioneers of resuscitation.[14] If suspended animation were real, it would require new physiological indicators, perhaps even new definitions of life and death, not just better diagnostic tests for the old vital functions .

The most radical view held suspended animation to be an interrup-tion, not just of life's indicators, but of life itself; a state in which "vitality is actually inert" ([213], pp. 19-28). In his pioneering 1745 essay on artificial respiration, Dr. John Fothergill equated reinflating the lungs with starting the pendulum on a stopped clock, to reanimate the "animal machine" ([79], p. 7; [230], p. 44). In this radically mechanistic explanation, suspended animation might be a curable form of death; resuscitation a form of resurrection ([169], p. 177; [222], p. 506).

The claim that life could be turned on and off like a machine directly challenged the prevailing vitalist theory and its insistence that living beings could never be reduced to purely mechanical processes. A typical early-nineteenth-century vitalist, Robley Dunglison, defined life as the ability to perform any functions which could not be explained by the laws of mechanics or chemistry. These inexplicable functions were attributed to the action of a "vital principle" – an immaterial essence of life often equated with the soul ([68], Vol. 1 p. 576; [17], p. 549).

True suspended animation posed a difficult challenge to such vitalist definitions of life and death. If the vital functions could be turned on and off like a clock, protested Edgar Allen Poe, "where, meantime, was the soul?" ([174], p. 258). Those vitalists who accepted the reality of suspended animation were forced to redefine life itself, abandoning the concept of vital *functions*, and emphasizing instead vital *capacities* or potentials.

Such potential life could be present without producing or requiring any physiologic activity at all. The absence of vital potential could never be directly demonstrated save by physical destruction of the body; diagnostic tests could confirm life, but could only infer death. Thus, redefining death as the loss of the capacity for vital functioning provided a way of reconciling vitalism with suspended animation ([55], p. 371; [110], p. 415).

The eighteenth-century Scottish physician William Cullen offered one important example of this new approach. Building on the neurophysiology of Haller, Cullen concluded that "nervous sensibility" and "muscular irritability" were the principles of life; perhaps the essence of life itself. "Life does not immediately cease upon the cessation of the action of the lungs and heart," Cullen wrote in 1774. "[T]he living state of animals does not consist in that alone, but especially depends upon a certain condition in the nerves and muscular fibres, by which they are sensible and irritable. . . . It is this condition, therefore, which may be properly called the vital principle in animals: and as long as this subsists . . ." life may be restored. Life was sensorimotor potential, not cardiopulmonary action ([44], p. 4).[15]

In addition to neuromuscular potential, another capacity of living beings that seemed unique and not reducible to eighteenth-century laws of mechanics was their power to resist the physical forces of entropy and decay. Thus, Georg Stahl concluded that a body's ability to stave off putrefaction proved it was still alive. Nineteenth-century vitalists like Marie-François-Xavier Bichat echoed Stahl, defining life as "the aggregate of the functions which resist death." For Stahl and Bichat, decomposition was not merely the only reliable *indicator* of death; it was close to being the actual *definition* of death ([68], Vol. 1, p. 576; [88], p. 312).

This somewhat circular vitalist definition retained support throughout the nineteenth century, though it was steadily constricted, as more and more processes formerly thought to be unique to living organisms were duplicated in the test tube ([88], pp. 473–474; [60], pp. 408-409).

Organs, Organisms, and Individuals: Physiology and the Levels of Death

Beyond their discoveries in artificial resuscitation and suspended animation, Enlightenment-era physiologists further complicated the concept of death with a series of experiments that raised troubling questions about the relation between the death of an individual and the death of his or her body parts. From antiquity, observers had been perplexed by the survival for a time of vital signs in the bodies of decapitated animals and people. The revival of anatomy and experimental vivisection in the seventeenth century lent a new urgency to such ancient concerns, as evidence accumulated that the heart, lungs, and brain could each function for a considerable time, in the absence of the others ([68], Vol. 2, p. 49). In fact, the first case of "brain death" to be maintained on a "respirator" may well have been the decapitated rooster, whose circulation and respiration William Harvey preserved with a bellows in 1627 ([14], p. 339).

If the vital organs could be maintained or terminated separately from each other, at what point in the process, if ever, does the organism-as-a-whole cease to live? And, at what point does its personal existence end?

Since the mid-seventeenth century, debate on these issues had been dominated by the essentially Platonic philosophy associated with the

work of René Descartes. In this Cartesian tradition, the human body
and the human self are totally distinct. Consciousness, pure mind,
comprises the self or soul – indivisible, immaterial, and immortal.
During life, the soul animates the body by means of the pineal body in
the brain. The body dies at the instant its connection to the soul is
broken. But the life of the soul remains distinct from and survives after
the death of all body parts. Only humans have conscious minds; non-
human organisms are soulless machines ([218], pp. 3–23; [70], p. 150;
[41]).

Between the mid-eighteenth and mid-nineteenth centuries the
findings of experimental physiology increasingly challenged key elements of this Cartesian tradition, and led even many vitalists to adopt
ever more materialist and more decentralized definitions of individual
life. For example, the Edinburgh physician Robert Whytt concluded, on
the basis of experiments with decapitated animals, that the soul and its
vital principles of sensibility and irritability were physically dispersed
throughout the whole nervous system, rather than being limited to a
non-physical relationship with a single organ like the pineal body ([88],
p. 326; [96], Vol. 2, pp. 68–72).

The French naturalist Georges Buffon further extended this decentralization of individual life. He postulated that all creatures were
composed of living "molecules," and that the life of an organism was
just the summation and structural organization of these separate molecular lives ([96], Vol. 2, pp. 5–17; [135], p. 166; [227], p. 256). The
eighteenth-century German physician Christian Hufeland carried these
conclusions to their logical extreme. For Hufeland, the death of an
organism was simply the progressive disintegration of its structure; not
an instantaneous event but a gradual process, ending only with complete
molecular decomposition ([2], p. 19; [227], p. 338).[16]

In an elegant series of animal experiments, the highly influential
Bichat attempted to prove that the organism was a combination of
separate vital functions performed by distinct bodily tissues. Bichat
distinguished what he termed "organic life" – the heart, lung, and other
functions necessary to maintain bodily metabolism – from "animal life"
– the functions of sensation and volition carried out in the brain [21,
224]. He further concluded that each of the body's twenty-one types of
tissue contained its own special vital principle, its own specific combination of sensibility and contractility ([88], p. 445).[17]

If there were different levels of life, each associated with different organs and different tissues, then an individual only some of whose tissues and organs had died was in a real sense part alive and part dead. Bichat observed that, after a stroke, "the patient may live internally for several days after he has ceased to exist beyond himself," a remarkably evocative description of what we might diagnose as the "locked-in syndrome" ([213], p. 11). His decentralized concept of the organism was extended even further with the mid-nineteenth-century promulgation of the cell theory ([96], Vol. 2, pp. 311ff.).

Not all early-nineteenth-century physiologists accepted the concept of individual death as a gradual, decentralized, and segmented process. Most continued to speak of the unitary, holistic, "somatic" life of the organism-as-a-whole, as distinguished from the "molecular" lives of its various component organs and tissues ([225], p. 791), [144], ([188], pp. 367–372).

Perhaps the most extreme challenge to the gradual nineteenth-century decentralization of individual life came from a Scottish physiologist, A. P. W. Philip. Philip singled out the powers of sensation and volition, lodged in the brain, as defining the life of the whole individual. "When the animal no longer feels and wills . . . he is, according to the common acceptation of the term, dead, although his body still retains its other powers," Philip declared. "[T]he changes which after this take place, of course no more affect the individual than if they took place in any other mass of matter" ([169], p. 170), [61–62].[18]

But defining individual life as consciousness and volition posed serious conceptual problems. How can anyone tell for sure whether someone else is capable of feeling ([167], pp. 162–164, 236)? The apparent privacy of other people's mental states left nineteenth-century physicians no practical way to operationalize a definition of death based on the absence of mental activity. And, by the mid-nineteenth century, growing interest in the existence of unconscious mental states precluded any simple Cartesian identity of consciousness with individual life ([27], p. 573), [69, 127].

The Premature Burial Panic: The Social Uses of Medical Definitions

Thus far, we have seen how new scientific discoveries in resuscitation, suspended animation, and experimental vivisection played a crucial and previously overlooked role in the Enlightenment-era wave of uncertainty over the diagnosis and definition of death. But such scientific developments are far from the whole story. For example, changes in the metaphors used to explain the organization of biological life (from the "monarchy of the heart" to the "republic of cells"), both affected and reflected revolutionary changes in the structure of social life ([146], p. 267). Scientific concepts of the relation between somatic and molecular death both influenced and responded to the development and use of the guillotine. And, the adoption of artificial respiration depended on the new bureaucratic structures, commercial expansion, and humanitarian values of eighteenth-century society [230].

Most importantly, diagnosing death is not just an abstract scientific problem; mistakes have drastic social and ethical consequences. Falsely diagnosing a living person as dead could result in withdrawing essential resuscitation, therapy, or palliation. Falsely diagnosing a dead person as alive could endanger the psychological, financial, or physical health of the family and community, while wasting medical resources. The types of mistakes a society worries about thus reflect a web of ethical, social, and political values. To fully understand a medical definition of life, it is necessary to understand the consequences it was intended to produce or avoid.

Between the 1740s and 1850s, most of the possible diagnostic errors evoked at least some expressions of concern. Most authors cautioned against false positive tests, such as mistaking the pulse in the doctor's fingertips for the pulse of the patient ([254], pp. 13, 121; [126], pp. 45–46; [209], p. 212). The motives for such warnings included both the professional fear of looking foolish ([254], p. 27), and the social fear that delay in burying the dead would spread disease ([254], p. 50; [17], p. 549; [92], p. xiii). This latter concern usually derived from explicitly utilitarian ethical values. Critics opposed saving a "single" individual from premature burial, if the cost would be to expose "thousands" to contagion from delayed interments ([103], p. 36).[19]

Like false positives, false negatives raised specifically professional concerns, such as the fear of malpractice suits ([213], p. 29), and more general ethical concerns, most notably that prematurely abandoning a patient was equivalent to murder. The latter opinion was defended in overtly anti-utilitarian arguments: to kill one patient was far worse than continuing to treat a thousand corpses ([254], p. 104).

By far the most typical and terrifying concern about diagnosing death was the desire to avoid premature burial. Edgar Allen Poe chillingly described "the unendurable oppression of the lungs – the stifling fumes of the damp earth – the clinging to the death garments –

the rigid embrace of the narrow house – the blackness of the absolute Night – the silence like a sea that overwhelms – the unseen but palpable presence of the Conquerer Worm – . . a degree of appalling and intolerable horror from which the most daring imagination must recoil" ([174], p. 263). What Poe left to the imagination, lesser productions, such as John Snart's *Thesaurus of Horror*, catalogued in lurid detail. The victim's final struggle, bathed in blood and excrement: "behold a *master-piece of horror!!!* – A torn and bloody shroud! battered forehead! broken knees and elbows! (O God!! O God !! *misericordia!!*)" ([215], p. 87; *see also* [122], pp. 171–173). Even the medical works on premature burial were sometimes graphically illustrated, such as Figure 1, a victim inadvertently rescued from the grave by a would-be graverobber ([254], opp. p. 7).

Across much of Europe, physicians, patients, and governments resorted to new and sometimes bizarre measures to prevent premature burial. Commonly, a minimum waiting period was required between diagnosing death and disposing of the body. By the early eighteenth century, church authorities in Italy and France instituted such delays of twelve to twenty-four hours ([227], pp. 398, 403; [10], p. 401). By the end of the century, governments began to impose much longer periods. Mantua and Tuscany had such laws by 1775 ([215], p. 81). The Code Napoleon required a twenty-four-hour delay throughout the Empire. Holland extended this to thirty-six hours, Austria to sixty-two hours, Frankfort and Saxony to three days ([227], pp. 398–406; [26], pp. 111–115; [123], p. 76).

Many jurisdictions also provided public mortuary buildings where these bodies could be kept for observation without posing too great a risk of contagion. The idea originated with Madame Necker in revolutionary Paris, and with Dr. Hufeland in 1795 Germany ([222], p. 493; [3], p. 28; [115], p. 17). It was first implemented in Frankfort in 1823, with Berlin, Paris, Vienna, Munich, and other central European cities following suit by mid-century ([123], pp. 75–76). Munich eventually built ten imposing establishments (*see* Figures 2 and 3) ([227], pp. 341, 345). Bodies were placed in flower-decked open caskets, with a bell rope in their hands to signal if they woke. The architecture permitted a centrally-stationed guard to observe most of the guests; twice hourly individual examinations were also to be made around the clock. The latest artificial resuscitation devices were always on hand ([227], pp. 338–365; [234], p. 226).

Fig. 1.   Rescued by a Grave-Robber, 1746. – National Library of Medicine.

Fig. 2. A Munich Mortuary, Circa, 1896. – National Library of Medicine.

Fig. 3. A Body in a Munich Mortuary, Circa 1896. – National Library of Medicine.

The most widely adopted legislation required some form of death certificate prior to burial. Such laws came in many varieties and served several different purposes. In nineteenth-century Britain and in most American jurisdictions, the primary purpose was to collect vital statistics. A physician might be required to fill out a certificate giving a medical cause of death, but need not have actually examined the patient ([227], pp. 282–289, 405; [65], pp. 51–53; [73], pp. 64–65.) Across the Continent, however, inspection of the body was central to death certification, with the prevention of premature burial a major stated objective. In different countries, private physicians, lay public officials, state-employed physicians, or some combination of these agents was assigned the task, with or without additional compensation, and with or without penalties for non-compliance. In Austria, Dalmatia, Saxony, and Munich, the laws spelled out the specific tests the certifying officer had to perform ([227], pp. 398–406). A variety of other purposes was tacked on to such laws, however, including inspection for signs of murder or abortion ([26], p. 115).

Since state-imposed measures were not always enforced or enforceable, especially in rural areas, many individuals took their own precautions. As early as 1662, French citizens began stipulating in their wills specific death tests and waiting periods to be completed before they were buried ([10], pp. 362, 397–399). Others designed their own escapable, alarmed, and provisioned coffins. By the 1840s, you could buy them ready-made, replete with signal bells and flags, speaking tubes, even an automatic ejection device.[20]

The diagnostic mistake worrying most people was not simply that they might be wrongly taken for dead, but was rather the specific fear of being *buried* alive. From the very start of the panic, and increasingly through the nineteenth century, doctors and lay people alike demanded bloodletting, embalming, autopsy, cremation, even decapitation and other mutilations, to guarantee that they would not be alive when buried.[21]

The specific fear of premature *burial* thus was more than simply a reflection of uncertainty about diagnosing death, but had many other sources. The epidemic diseases that swept eighteenth- and nineteenth-century cities led to a public health movement, one of whose reform proposals involved prompt burial of the dead in new remote rural cemeteries. These reforms interfered with traditional ethnic and religious mourning practices, such as family wakes, and sometimes provoked riots when families sought forcibly to retain their dead at home. Hasty burials in far-away places contrary to tradition undoubtedly fed the fear of premature interment. The new burial practices also fed the related fear of grave-robbery, as the lower classes suspected the reforms were designed to make it easier for doctors to obtain fresh cadavers for medical schools ([254], title page; [150], pp. 104–106).

But many of those who feared premature burial were not from among the urban poor. For the leaders of the panic, the intellectual trends of the Enlightenment may have been more important, from a new sensitivity to individual suffering ([167], Chapters 3, 5, 11), to the emotional vacuum left by secular rationalism when confronting death without heaven or hell. Premature burial plausibly may have been a secular equivalent to hell for those who no longer found the original quite so terrifying ([2], p. 20).

Because Jewish ritual demanded prompt interment, the attack on premature burial also gained support from Europe's pervasive anti-Semitism. Although the early authors mentioned Jewish burial practices only to praise the ancient Hebrew custom of reinspecting their crypts ([254], pp. 156–157), once governments attempted to impose mandatory waiting periods, contemporary Judaism came in for harsh criticism, led by sensationalist fear-mongers like John Snart. "[S]uch was the inflexible attachment of the Jews to their ceremonials," he wrote, "that they became the continual subject of execration, even of CHRIST himself . . .; and they have never reformed since!" Calling premature burial "the barbarous custom . . . which the Jews . . . are notorious for doing," he ominously demanded "removal of the premature Jewish custom . . . throughout the habitable globe." However, Snart reserved his most splenetic attack for Catholics, because of the Church's opposition to rational science, such as he ostensibly represented ([215], pp. 6, 37, 93–94). In fact, the Jewish community itself briefly split over the issue. Leaders of the Jewish Enlightenment (*Haskala*), such as Moses Mendelssohn, urged compliance with the new laws in the name of science and rational humanism ([4], pp. 287–295).[22]

There were also clear national differences. Fear of premature burial always seemed stronger in France and Germany than in England or America ([76, p. 205; [123], p. 71; [170], p. 31). Echoes of the difference still persist. Recently Oxford University Press published John McManners' study of *Death and the Enlightenment*, a subject the eminent French historian Philippe Ariès also treated as part of a general history of death. Ariès devoted over sixty pages to premature burial, including lengthy stories he remembered hearing in childhood; McManners dismissed the phenomenon in a paragraph, calling it a "bizarre" manifestation of the French "hypochondriacal zeal" ([10], pp. 353–404, passim; [135], pp. 47–48).

Fear of premature burial did feed upon more general psychological terrors. Edgar Allen Poe's images of burial alive emphasized a claustrophobic dread of powerlessness and confinement, "the rigid embrace of the narrow house" ([174], p. 263). In fact, Poe's tales of living people imprisoned in the grave mirrored his equally terrifying depiction of "M. Valdemar," a man whose death agony was prolonged for weeks when his soul was trapped within his already dead body by means of mesmerism [175]. Thus, while physicians' uncertainties over defining and diagnosing death played a major role in the premature burial panic, many other social forces contributed to the specific fear of being buried alive.

III. LATE-NINETEENTH-CENTURY DEATH: FAITH IN EXPERTS, LINGERING DOUBTS

During the late nineteenth century, doubts about doctors' ability to define and diagnose death gradually subsided, due more to growing public faith in medical expertise than to new technical discoveries. By the end of the century, such issues had almost disappeared as topics of legitimate professional concern. Yet, scientific uncertainties about death, and popular fears of premature burial, remained strong far longer than most previous historians have recognized. The scientific unknowns continued to result from new work in resuscitation, suspended animation, and experimental vivisection. But the fear of premature burial was increasingly relegated to those segments of the population that opposed the growing social power and increasingly technical content of professional medicine.

The New Professional Certainty

Late-nineteenth-century physicians demonstrated greatly increased self-confidence in their ability to recognize death. Their new willingness to shake off the doubts of the previous century was based in part on the introduction of precise new technical instruments to test for vital functions [189].

René Laennec invented the most important of these tools, the stethoscope, in 1819. Another French physician and medical instrument pioneer, Eugene Bouchut, first applied the new device to the systematic diagnosis of death in 1846, an innovation for which he received two major prizes ([177], p. 93), [211], ([63], Vol. 6, p. 1242). By the third edition of Bouchut's treatise, in 1883, both the stethoscope itself and doctors' skills in its use had improved so much that many physicians felt sure they could detect even the faintest traces of heart or lung activity [24], ([187], p. 146; [107], p. 562; [99], p. 88).

Electrical tests for neuromuscular functions also became more technical and mechanized, with the invention of new devices to produce, control, and measure electrical currents [201], ([203], p. 370; [137], p. 18). Such tests won an additional two of the prizes awarded for new ways of diagnosing death [211], ([227], p. 424). In 1920, *Scientific American Monthly* labelled the failure to respond to electrical stimulation an "infallible" indicator of death [75].[23]

Carl Wunderlich first popularized medical measurement of body temperatures in the 1860s ([1], pp. 108, 159). Between 1868 and 1880, four Britons and a Frenchman won prizes for the use of thermometers to check for continued metabolic functioning, "vital fire," as a test for death ([192], p. 107; [187], p. 146), [211], ([227], p. 424). Other new mechanical death tests included: high-intensity lamps to examine the skin between the fingers for signs of circulation ([192], p. 113; vs. [194]); microphones to amplify chest sounds ([107], p. 564); and x-ray fluoroscopy to search for motion of the internal organs ([36], p. 260).

Perhaps because of the eyes' reputation as "windows of the soul," many new instruments were used to examine the eyes for signs of death. The ophthalmoscope made possible examination of the retina and sclerotic for lingering signs of circulation and early signs of decomposition ([177], pp. 94–95; [26], Chapter 3). Time-lapse photography could record pupillary reflex irritability in response to stimulation with such new drugs as atropine and belladonna ([8], p. 107).

Mid-nineteenth-century improvements in the hypodermic syringe and in organic chemistry led to such additional tests as the subcutaneous injection of ammonia (for evidence of inflammatory responses) ([192], p. 114; [177], p. 92), and the intravenous injection of flourescin ($C_{20}H_{12}O_5$). According to S. Icard, who won two prizes for devising the latter procedure, if any circulatory and metabolic activity remained, fluorescin would turn the eyeballs bright green ([238], p. 42; [137], p. 18; [161], p. 18).

Late-nineteenth-century death tests were not only more technical but also more invasive than their predecessors. Surgeon Jules-Germain Cloquet and artificial respiration innovator J.-B.-V. Laborde claimed that a shiny steel needle inserted deep into living muscle would be metabolized and rust, while a needle stuck in a dead muscle would remain untarnished ([192], p. 114; [177], p. 92). To test for an inflammatory response, various doctors burnt the skin with open flames, boiling water, or "deep insertion" of a heated cautery ([137], p. 18; [177], p. 92). A Dr. Josat invented a "nipple pincher"; while a Dr. Middeldorpf tested for cardiac motion by jabbing a long needle with a flag on one end directly into the heart ([26], pp. 51, 58).

These new tests supplemented rather than supplanted earlier methods. Some older death signs, such as livid patches and flaccid corneas, actually gained increased acceptance over the nineteenth century.[24] Even the use of mirrors to test for breath still had medical advocates as late as 1882 ([177], p. 92). Opening a blood vessel to test for circulation remained the favorite technique of the flamboyant forensic pathologist Sir Bernard Spilsbury in 1930 [157, 61].

Rather than accept any one test as definitive by itself, many physicians concluded that combining the best fifteen or twenty such procedures would adequately insure that vital functioning had in fact ceased. This shotgun approach, which originated with toxicologist Matthieu Orfila in 1818, became established as professional orthodoxy by the 1880s ([158], p. 154; [177], p. 95; [3], p. 30).

Unresolved Scientific Difficulties

Medical confidence in these new diagnostic tests grew in spite of, rather than because of, new scientific knowledge about death in the late nineteenth century. New discoveries in experimental physiology continued to raise potentially troubling questions for doctors about the definition and diagnosis of death.

To accept the reliability of the new tests, physicians had to deny the possibility of true suspended animation. Doctors now insisted that previously reported recoveries from apparent death resulted from the insensitivity of earlier diagnostic tools, rather than from any actual intermission in the heartbeat and breathing. "[T]he vital functions, although they fall to a low ebb, are never altogether suspended," a British physician asserted in 1898 ([238], p. 6). An American doctor agreed that, "Like the respiratory movements, the movements of the

heart may be very slow and infrequent," but that "[c]ontinuous cessation of the cardiac function precludes all possibility of a return to life; it
may be regarded unequivocally as indicating death. . . ." ([107],
p. 559).[25]

But late-nineteenth-century physiological discoveries hardly justified
such medical confidence. For example, in 1874 Moritz Schiff used open
chest cardiac massage to revive a heart that had stopped beating. Unlike
eighteenth-century techniques for restoring the circulation, open chest
heart massage provided direct, dramatic evidence that true cardiac
arrest did not mean instant or certain death ([164], p. 5; [26], pp. 20, 29;
[36], p. 260). Other, similarly problematic resuscitation methods pioneered by researchers in this period included closed chest heart massage, adrenal extract injections, and a variety of primitive mechanical
respirators ([164], pp. 8, 14; [14], pp. 344, 347; [2], p. 23; [148], pp.
3–10).

By the 1880s and 1890s, these and other improved resuscitation
techniques occasionally produced instances in which lung and heart
functions were artificially maintained for many hours, following extensive brain damage. Many modern physicians probably would diagnose
such cases as "brain dead," and regard them as casting doubt on the
value of heart-lung function death tests. But late-nineteenth-century
physicians saw in these patients nothing more than an unusually prolonged confirmation of the ancient doctrine that the heart was the last
organ to die [66, 172].[26]

Late-nineteenth-century advances in resuscitation were often made
both possible and necessary by the development of anesthesia – especially with the 1848 introduction of chloroform, a drug which not
infrequently triggered sudden cardiac arrest ([167], Chapter 3; [164],
pp. 5, 8, 10, 29; [203], pp. 370–371). These accidents, along with the
intriguing similarities between anesthesia and suspended animation,[27]
gave anesthetists special reasons to study death and resuscitation. Anesthesiologists have figured prominently in efforts to define death ever
since [191–192], [184], pp. 62–72; [148], p. 21).

Anesthesia pioneer Sir Benjamin Ward Richardson dominated late-
nineteenth-century research on suspended animation. In 1879, he
startled both lay and medical audiences with his claim that drugs and
cold could produce suspended animation in humans, and that he was on
the verge of discovering how to revive such patients after "weeks and
possibly months." "[I]t may be doubted whether a healthy, warm-

blooded animal, suddenly and equally frozen through all its parts, is dead," although we have not yet perfected the means to safely thaw it, he explained ([191], pp. 100-101), [183], ([192], pp. 117; [8], p. 106).[28]

Richardson and other physiologists who believed in suspended animation emphasized purely material causes, such as specific drugs, diseases, and physical conditions. Although they recognized that psychologically disturbed individuals, particularly young women, seemed especially susceptible to suspension of the vital signs,[29] late-nineteenth-century researchers either ignored or traced to physical causes such previously postulated psychogenic death-like states as catalepsy, ecstasy, and trance. Richardson drew up a long list of drugs capable of producing apparent death, including: atropine, amyl nitrate, belladonna, curare, chloral hydrate, hydrogen cyanide, and pure oxygen ([191], p. 100). [30] To the anesthetist Richardson, catalepsy was simply the result of endogenously produced amyl nitrate ([191], p. 100). To neurologists like George Beard and William Hammond, trance, ecstasy, and other allegedly "spiritual" causes of suspended vital signs were really lesions of the brain or spine ([27], p. 572). These researchers thus attempted to stake out a professional territory, midway between the majority of clinicians, who denied the very possibility of suspended animation, and the lay public, whose interest in the "occult" dimensions of the phenomenon continued unabated [27, 208], ([10], p. 457; [100]).

Organs, Organisms, and Persons: The Conflict Over the Integration of the Individual

Some of the most difficult questions about death grew out of the perennial problem of distinguishing the deaths of an individual, an organism, and its various body parts. Even as physicians renewed their confidence in heart, lung, and muscle function tests to diagnose individual death, many physiologists had begun to conclude that individual life depended instead on the nervous system, while others continued to reject the entire concept of individual death as distinct from the death of the separate body components.

Late-nineteenth-century vivisectionists employed increasingly delicate surgical dissections in a new and determined effort to isolate the specific functions of each specific organ and structure. In 1880, Sidney Ringer showed that a heart, removed from a frog, could be maintained alive in a salt solution. His demonstration, and the earlier perfusion experiments of Carl Ludwig, made it necessary to inquire into the status of an organism, all of whose vital organs were kept functioning, separately, *in vitro* ([88], pp. 556, 560; [61]).

Many physiologists answered that organisms had no individual lives, apart from the vitality of their component tissues and cells. The eminent American Austin Flint Jr. denounced as "unphilosophical" the effort to define the moment at which the life of the organism-as-a-whole terminates. "There seems to be no such thing as death, except as the various tissues and organs which go to make up the entire body . . . lose their physiological properties . . .; and this occurs successively and at different times for different tissues and organs." His position was the "commonly accepted explanation" among "the best modern physiologists" in 1891, according to a New York forensic pathologist ([144], pp. 175–176; [74], p. 54).

While Flint's conclusion clearly derived from the earlier work of Bichat, it gained much additional support from nineteenth-century cell theory, in which all multicellular organisms could be portrayed as a commonwealth of independent living monads ([96], vol. 2, pp. 179–209; [222], p. 497; [160], vol. 1, p. 1834). Further support came from intriguing new findings in plant physiology. Discovery of the alternation of generations, in particular the startling, part multicellular, part unicellular life cycle of the slime molds, made the very existence of discrete individual organisms seem merely an illusion ([60], p. 12), [121], ([229], pp. 6–9, 14). In 1897, the pathologist of the Paris morgue concluded, "The unity of the living being is therefore only apparent . . ." ([26], p. 25).

Yet, if the unified life of the individual was merely an illusion, it proved an unusually hard one to dispel. Even the Paris pathologist just quoted found the concept inescapable. Describing a guillotined convict, he wrote that

One hour after the execution the heart still beat; yet this man's existence was over; he had lost his personality, and yet his heart was beating! Well, to us and to everyone a decapitated person is a dead man, although his heart does continue to contract ([26], p. 20).

What was this "personality" whose loss differentiated a dead individual from his still living body parts? To the conceptual founder of modern neurophysiology, Charles Scott Sherrington, what distinguished the life of an organism from the lives of its parts was its ability to integrate and coordinate those parts, a capability vertebrates exercised through their nerves. The nervous system for Sherrington was far more than simply a transmitter of sensation and motion; it made possible the integrated functioning that defined a living individual [207], ([88], p. 543; [70], pp. 141–142).

Sherrington's concepts developed in the context of the bitter debate over whether the functions of the brain and nerves were localized in specific neurologic structures or were performed by the nervous system as a unit. With the localization of specific mental abilities by Broca and Ferrier, and the isolation of specific neurological regulators for respiration and for heartbeats, the study of specialized sub-units appeared to have triumphed over the holistic approach advocated by Frederick Goltz ([60], pp. 62–63, 178, 270; [88], pp. 466–467, 540–541). Sherrington reasserted the unity of the individual organism and the interconnectedness of the nervous system ([60], p. 63).

Is an individual then nothing but a functioning nervous system? Unlike such reductionists as Emil du Bois-Raymond and Hermann Helmholtz, Sherrington and John Hughlings Jackson viewed the self and the nerves as parallel but separate. For Jackson and Sherrington, each activity of the nerves correlated with a corresponding mental state, but mind and body remained distinct. The nerves made the integration of the individual possible; the individual, however, was not identical with the nerves ([60], pp. 246, 271), [70, 257], ([96], Vol. 2, pp. 272–280).

The Social Roots of Medical Certainty

The uncertainties created by physiology contrast starkly with late-nineteenth-century physicians' confidence in their ability to diagnose death. Although the innovations in diagnostic technology played a role, the new medical self-assurance owed more to social forces than to technical discoveries.

Physicians candidly acknowledged that their increasingly complex combinations of highly technical and invasive death tests won medical acceptance in large part because they were too complicated for lay people to perform. The tests appealed to physicians because they helped the medical profession gain a new monopoly over the diagnosis of death. In 1870, the founder of the Paris morgue speculated, "Suppose for a moment that a sign of death obvious to the whole world were discovered. What would happen? *Persons wouldn't call a physician to verify death.* . . . Let us therefore abandon the unfortunate idea that the signs of death must be vulgarized" ([3], p. 30; [26], p. 35; [211]).

The complexity of the new diagnostic tests enabled the newly emergent medical profession to offer the public a carrot and a stick. A medical monopoly would promise reliable death diagnoses; any other arrangement would insure the continued fear of premature burial. The director of the Paris morgue in 1897 insisted "that the uncertainties which sometimes arise at the moment of death, and in the hours which directly follow, may be removed by the physician; that his intervention is often necessary to establish the reality of a person's death, and that this alone can remove every apprehension . . . of premature burial" ([26], p. xi; *see also* p. 50; [3], p. 30).[31] In strikingly derivative fashion, early-twentieth-century "funeral directors" also claimed that their new professional skills were necessary and sufficient to protect the public from premature burial ([74], pp. 162–163).

Most advocates of professional monopoly relied on the complexity and supposed accuracy of their new technical instruments. However, a few also justified their expertise by appealing to the older arts of clinical observation and experience. "Death," one German physician maintained, "is best recognized by the total impression it makes on us" – an explanation remarkably reminiscent of Justice Potter Stewart's well-known admission that, while he might not be able to define pornography, "I know it when I see it" (*Jacobellis v. Ohio*, 378 U.S. 184, at 197 (1964)) ([75], pp. 396–397; [170], p. 31).

Physicians who portrayed medical death certification as a preventive against premature burial sought more than simply the meager additional income from the inspection of corpses ([22], p. 308; [238], p. 11). The power to determine death would also give doctors the power to set the boundaries around life's most final and fearsome mystery, and thus would enhance, at least symbolically, the status of the medical profession ([10], p. 403; [74], p. 52).

A surprising range of other professional interests was also implicated. Cemetery reformers and public health officials, eager to stamp out unhealthy traditional mourning customs, used the fear of burial alive to win public compliance with new sanitary burial practices, and to increase public support for the medically supervised mortuaries ([10], p. 402; [74], Chapter 4; [123], pp. 75–76; [227], p. 342). John Shaw Billings, who sought medical death certification laws to improve the collection of vital statistics in the United States, argued that such legislation would also help to prevent premature burials ([22], p. 308). Others claimed mandatory medical inspection of the dead would improve rural access to medical care ([26], p. 114).

Concern that the sick and unwanted might be disposed of by prematurely declaring them dead surfaced as early as the 1890s. With the emergence of organizations advocating euthanasia and eugenics, such fears soon became deeply entwined with the issue of defining death. In 1895, a British periodical warned that the world was full of "superfluous people" – the elderly, infirm, and poor – whom relatives and lay authorities might not examine too carefully, and who thus might be buried alive. According to the author, medical inspection of the dead would solve this problem, too ([31], p. 332).

As medical opposition to abortion crystallized in the late nineteenth century, French physicians also began using death inspection laws to uncover hidden evidence of feticide ([26], p. 115). Defining when "personhood" ends is logically similar to defining when it begins; to that extent, the definition of death and the abortion controversies shared a common focus. However, in the nineteenth century these two questions only rarely coalesced. There were crucial emotional differences between declaring the end of pre-existing personhood in a familiar adult, and attributing new personhood to an unseen, unknown fetus ([91], p. 97). Furthermore, the specific nineteenth-century obsession with preventing premature burial, by the use of methods such as cremation and decapitation, precluded any nineteenth-century "right to life" coalition from uniting the two issues.

The movement for death certification thus linked the definition of death with a broad spectrum of other medical issues. In some cases, physicians exploited public fears of premature burial, in an effort to medicalize death for their own professional advantage. However, many doctors genuinely believed that a medical monopoly on diagnosing death would help save the public **from a** real and present danger of being buried alive ([3], p. 29).

Furthermore, professional efforts to co-opt and take advantage of the premature burial panic could be a two-edged sword. Those who feared living interment attacked everyone involved in diagnosing death. To the extent that doctors took over this function, they incurred increased portions of the blame. And, the undercurrent of irrational terror and credulity always lurking beneath the surface of even the most scientific works on the subject increasingly was seen as incompatible with the new medical professionalism of the late nineteenth century ([215], pp. 77–78; [225], p. 803; [161], p. 23; [100]).

Thus, many physicians shunned any attempt to ride the wave of premature burial panic. Instead, they began to deride the entire subject, as an irrational fear of an impossible event.

[I]n spite of the rank harvest of literature that has sprung up around the subject, it may be at once stated that the whole theory of premature burial is unsupported by a single scientifically proved instance; that the likelihood of such an occurrence is extremely small; . . . in short, that the whole of this popular belief is nothing more than a legend,

declared a British physician in 1898 ([238], p. 5; [170], p. 34; [179]). Doctors investigated published accounts of premature burials and declared them fraudulent [15], ([238], p. 44), or subject to alternative explanations. The sounds of movement heard inside coffins, the positional changes found in disinterred corpses, were attributed to vandalism, decomposition, rats, and other natural phenomena, rather than to the supposed struggles of the deceased ([26], p. 32; [99], p. 100; [137], p. 17).

While some professional leaders did use public belief in premature burial to attack irregular healers and lay coroners as dangerously incompetent, the medical majority, composed of ordinary general practitioners, feared that elite specialists could use the same argument against them.

[T]ho many of our leading physicians . . . [believe] that a proportion of our 'dead' are interred prematurely, the rank and file of the profession have as systematically pooh-poohed the idea, and that for rather obvious reasons. To begin with, they do not wish it to be supposed that they cannot distinguish between life and death,

according to a 1907 believer in the danger of burial alive ([95], p. 682; [205], p. 530: [107], p. 555).

Most fundamentally, the new confidence in doctors' ability to diagnose death reflected an unprecedented general faith in professional expertise. This popular enthusiasm for science and scientists undoubtedly reflected the tremendous technical advances of the era, including the revolutionary medical discoveries in bacteriology and immunology made by Pasteur, Koch, and their followers. But popular enthusiasm for science grew even more rapidly than did technological accomplishments. Throughout turn-of-the-century Western Europe and the Americas, a variety of "progressive" social movements built upon belief in the power, objectivity, and orderly efficiency of science, to establish an unprecedented degree of trust in and deference to technical experts [247, 220].

In this regard, late-nineteenth-century death tests bear a striking similarity to early-twentieth-century I.Q. tests. The death tests arguably had a firmer scientific basis. But the professional and public confidence in each derived far more from the new deference to technical expertise, than from a careful analysis of the underlying evidence [93].

The Organized Movement to Prevent Premature Burial: 1866 to the 1930s

Despite the era's unprecedented faith in experts, the late nineteenth century produced a significant, if shortlived, organized movement to combat premature burial. The seeds of this campaign were planted in two separate appeals to legislative bodies: an 1866 petition to the French Senate by Ferdinand Cardinal Donnet of Bordeaux, and an 1871 lecture before the New York State Assembly by Eclectic physician Alexander Wilder ([8], p. 104; [227], p. 113; [248]). Both stressed the inability of medical experts to diagnose death with certainty, and the need for legal action to assure that the average physician met even the imperfect performance standards of the profession ([249], p. 20). Their legislative proposals were defeated ([154], p. 572), and little more was heard of them, until 1895.

In that year, William Tebb of London edited a pamphlet excerpted from Wilder's address, under the title *Perils of Premature Burial*. Wilder's speech also figured prominently in Austrian Franz Hartmann's *Buried Alive*, the book that directly launched the new movement [249, 100]. Though satiric suggestions for a "Society for the Suppression of Premature Interment" had appeared as early as 1819 ([228], p. 335), Hartmann's book comprised an unprecedented plea for the actual creation of such associations.

The following year, Tebb, James R. Williamson, and Hartmann's American friend Col. Edward Perry Vollum, organized the group known as either the Association or the Society for the Prevention of Premature Burial (SPPMB) [95], ([101], Chapter 3; [227], p. 18). Concurrently, Tebb and Vollum published *Premature Burial and How It May Be Prevented* [226]. Extensive press coverage of the SPPMB, partly orchestrated by the members themselves, evoked an outburst of public concern. A single article in *Le Figaro* elicited over 400 letters to the editor ([95], p. 681).[32] Historians who argue that the premature burial panic ended in the mid-nineteenth century base their conclusions on the optimistic claims of the era's physicians, not on the extensive lay literature ([3], p. 31; [122], p. 174; [180], pp. 15, 82).

Among its many activities, the SPPMB promoted the famous coffin signal device invented by Russian Count Michel de Karnice-Karnicki, Chamberlain to the Czar. This mechanism, patented in Berlin in 1897, was triggered by a glass ball placed directly on the buried person's chest. Any movement of the ball activated a spring trigger, which in turn opened an air duct, turned on an electric lamp, rang a bell for half an hour, and raised a flag four feet above ground level. It sold in England for only twelve shillings ([238], p. 42; [227], pp. 320–322; [137], p. 19; [11], p. 1952).

The SPPMB also urged British and American legislatures to adopt the full range of Continental death legislation ([227], esp. pp. 284–289; [101], p. 127; [249], pp. 8, 13). For example, a bill introduced in the New York State Assembly in 1898 would have required that a doctor or coroner certify all deaths, based on five specified death tests; mandated construction of public mortuaries, with full-time medical staff; and imposed a seventy-two-hour waiting period before burial [38], ([155], Vol. 2, p. 1025). A similar bill placed before both houses of the Massachusetts General Court in 1904 proposed that doctors be fined if they abandoned resuscitation efforts prior to the appearance of a total of eleven specified signs of death [179].

Despite these efforts to legislate specific lists of death tests, the SPPMB did not expect such procedures alone would prevent premature burials. On the contrary, they retained the firm conviction that only putrefaction provided a foolproof diagnosis.[33]

The eighteenth-century premature burial panic had originated with new medical and scientific discoveries, but medical innovation was the target, not the source, of the SPPMB campaign. Though they continued to draw upon eighteenth-century medical authorities and concepts, the SPPMB ignored or denounced most late-nineteenth-century scientific discoveries, even those that could have lent considerable support to their efforts. All five death tests required by New York's proposed anti-premature burial statute were long outdated by 1898. They included mandating the use of a mirror to test for respiration; the stethoscope was not even mentioned! ([38], pp. 5–7). And, while medical science increasingly required quantified clinical studies and controlled laboratory experiments, the works of Hartmann, Tebb, and Vollum continued to rely almost exclusively on the accumulation of disparate anecdotal accounts, little changed in style or substance from the works of eighteenth-century physicians like Bruhier ([101], esp. pp. 1–63; [227]).

The discoveries of such nineteenth-century physiologists as Schiff, Flint, and Ringer could have provided useful ammunition to attack the conceptual and operational weaknesses of medical death tests. Yet, with the exception of some of Richardson's work on suspended animation,[34] the SPPMB virtually ignored contemporary physiology. This omission almost certainly resulted from the Society's close ties to the antivivisection movement. Surgeon Walter Hadwen, who edited the expanded second edition of Tebb and Vollum's *Premature Burial*, served as head of the Society for the Protection of Animals Liable to Vivisection. The founder of that group, Frances Power Cobbe, also joined in the anti-premature burial agitation, as did George T. Angell, the founder of the Massachusetts Society for the Prevention of Cruelty to Animals ([233], pp. 119, 176n67; [227], p. 306).

The SPPMB and the antivivisectionists were drawn together by more than a shared antipathy for suffering. They each formed a link in a chain of organizations opposed to the scientific revolution in late-nineteenth-century medicine, and to the growing social power of doctors. In addition to anti-premature burial and antivivisection, this coalition included antivaccination, antilicensure, antibacteriology, medical sectarianism, occultism, and spiritualism. Each of these movements professed allegiance to an older model of medical science, in which cleanliness and godliness, not laboratories, germs, and vaccines, held the keys to human health. And each appealed to an older libertarian political philosophy, which portrayed professional power as a fundamental threat to personal freedom [83, 120, 165, 221, 233].

The anti-premature burial leadership fully reflected this wider critique of medical science and medical authority. Alexander Wilder (1823–1908), the ideological godfather of the SPPMB, founded, and for twenty years headed, the National Eclectic Medical Society. One of the foremost spokesmen for American sectarian medicine, Wilder won official Eclectic endorsement for anti-premature burial legislation. He also authored *Vaccination, a Medical Failure* in 1875, and, with William Tebb, founded America's first national antivaccination organization ([120], p. 465; [227], p. 408).

To the end of his long career, Wilder completely rejected the bacterial origin of disease. Suspended animation, and all other diseases, he attributed to the environmental and moral etiologic agents emphasized by pre-bacteriological health reformers: "surfeiting, undue exposure to heat or cold, over-taxing of the brain, excessive bathing, mental excitement, sexual aberrations, the use of tobacco or anaesthetics, unwholesome postures in bed, etc." ([249], p. 20). His life-long interest in the occult also led him to produce such works as *The Secret of Immortality Revealed* (1846), and *The Antecedent Life* (1896), and to edit several volumes on Hinduism and serpent-worship ([131], p. 8; [165], pp. 153, 303; [202], pp. 221, 279).

SPPMB cofounder William Tebb crisscrossed the Atlantic to help create pioneer antivaccination societies in New York (1879) and London (1881). An inveterate traveler and organizer, born in 1830 in Manchester, England, Tebb lived in Massachusetts, London, and India ([120], p. 465; [165], p. 99; [227], p. 19). Walter Hadwen combined his efforts against premature burial and vivisection with his "furious" opposition to vaccination ([233], p. 119). Franz Hartmann's pursuits spanned the spectrum of the supernatural, from magic to Rosicrucianism; his *Buried Alive* was first published by the "Occult Publishing Company" of Boston [100]. Arthur Lovell, a secretary of the SPPMB, was the promoter of the "Ars Vivendi" system of unorthodox healing, and author of self-help health guides that stressed the value of deep breathing as a panacea ([238], p. 44).[35]

The SPPMB, even more than its eighteenth-century predecessors, focussed almost exclusively on avoiding premature *burial*, not on other errors in diagnosing death. The title page of Tebb and Vollum's *Premature Burial* carried an aphorism from Wilder, "The thought of suffocation in a coffin is more terrible than that of torture on the rack, or burning at the stake" [227]. "It would be better to cut the carotid artery, or put a steel needle through the heart, even at the risk of killing a few entranced subjects, than to bury alive one man . . .," asserted another SPPMB supporter [31].[36] Hartmann's proposal to require that all coffins be filled with chloroform seems the most humane of the suggested preventive measures ([101], p. 127). SPPMB leaders and sympathizers specified in their wills that such drastic steps be taken before they were buried, and most of these requests seemingly were carried out ([95], p. 681; [137], p. 19: [11], p. 1951; [74], p. 151).

SPPMB rhetoric, like the earlier stories of Poe, emphasized claustrophobic feelings of wrongful confinement and powerlessness, emotions that help explain why many anti-premature burial leaders also joined other social movements to abolish everything confining, from Negro slavery to mental institutions. Wilder, Tebb, and other SPPMB figures began their reform careers in the fight against slavery, particularly the immediatist, anti-institutional Garrisonian branch of abolitionism. And, anti-slavery continued to provide an important metaphor for their later crusades against the power of doctors ([131], p. 8; [227], p. 19; [120], p. 466; [165], pp. 99, 188, 262). (One of the earliest campaigners to free women from the unhealthy confinement of tight corsets was the anti-premature burial pioneer J. B. Winslow [62]).

Alexander Wilder's involvement in the movement to prohibit involuntary psychiatric commitments is especially revealing of this claustrophobic impetus to reform. Wilder's seminal 1871 legislative remarks on wrongful burial were actually part of a longer address, devoted primarily to wrongful mental hospitalization. In fact, Wilder denounced involuntary commitment as "burial in the living tomb," and warned that, "upon the certificate of a single physician," any one could be "committed to the cell of a lunatic asylum, the walls of which are as high and strong, the keepers as vigilant and morose, the . . . doors as difficult to escape from," as those of a prison or tomb ([248], pp. 35, 37). In both cases, the conceptual inadequacies of materialist science and the unchecked power of the medical profession consigned innocent victims to hopeless confinement (*see also* [133]).[37]

From sectarians and libertarians fighting for an unregulated medical marketplace, to opponents of involuntary hospitalization and vaccination, the image of doctors burying people alive provided an evocative metaphor for all those who felt restricted and confined by organized medicine. Furthermore, those who saw the new technological, laboratory-based, materialist, and amoral medical science as a betrayal of older holistic health ideals, welcomed the SPPMB message that, despite their new technology, doctors still could not even tell the living from the dead.

Did the fear of premature burial have a rational basis? Were people "really" being buried alive? Without doubt, some mistaken diagnoses of death were made, and individuals almost certainly did revive after having been pronounced dead. Leaders of the anti-premature burial movement, from Winslow to Cardinal Donnet and Col. Vollum, insisted that they themselves had narrowly escaped burial during terrifying bouts of suspended animation ([101], p. iii; [8], p. 104). Others witnessed similar close calls ([137], p. 19).

But did anyone ever regain consciousness inside a buried coffin? Virtually all of the evidence available can be explained in other ways. Yet, it would be very difficult for anyone who actually experienced true living entombment to return and provide definitive evidence. To that extent, premature burial, like the afterlife, is an inherently untestable, i.e. non-falsifiable, hypothesis. It is precisely this inability to know for certain before it is too late that explains the terrifying fascination of both premature burial and hell.

The SPPMB succeeded in reviving the premature burial issue by linking it to opposition to medical science and hostility to the power of doctors. But, as popular adulation of scientific expertise reached a peak in the early decades of this century, the fear of burial alive gradually receded. In the English-speaking countries, virtually all of the SPPMB's legislative proposals met total defeat. The British Burial Act of 1900 still permitted doctors to sign death certificates without their having examined the patient ([227], p. 289). As late as 1932, some areas of the United States lacked any formal death registration system ([71], p. 23). Although Central Europe continued to add to its anti-premature burial legislation, the only American jurisdiction known to have adopted such laws was Dover, New Hampshire, in 1895 ([227], pp. 244–247, 293–299, 308, 401).

Concern about premature burial, however, never completely disappeared. The SPPMB continued to be active at least through 1923 ([115], p. 17). English language articles on the subject were listed in the *Readers' Guide to Periodical Literature* through 1920; French and Italian medical articles appeared in the Surgeon-General's *Index-Catalogue* through the 1930s (*see also* [115], p. 33); and evidence of concern in those countries can still be found today [168], ([10], p. 397; [180], pp. 152–153). As late as 1940, an article in *Scientific American* claimed that "frequent" errors in diagnosing death were still causing cases of premature burial ([156], p. 337). Faith in medical expertise suppressed public fears of premature burial, without ever fully eradicating them.

IV. THE TWENTIETH CENTURY: A FASCINATION WITH UNCERTAINTY

Even as the public's fear of premature burial receded into the shadows, new discoveries in medicine and physiology continued to provoke potentially troubling questions about the meaning and recognition of death, throughout the twentieth century. These disturbing findings came in the same fields that had undermined eighteenth-century distinctions between life and death: resuscitation, suspended animation, and the physiological relation between the individual organism and its component parts. Yet, prior to the late 1960s, public and professional discussion of this work focussed on the marvels, rather than the dangers, to be expected from the inadequacy of our definitions of death.

The man who revolutionized our conception of the relationship between an organism and its parts was America's first medical Nobel Laureate, Alexis Carrel. Building on the work of Ludwig, Ringer, and Ross G. Harrison, Carrel, between 1910 and 1920, pioneered the culture of cells, tissues, and entire organ systems in artificial media outside of the body. He claimed (falsely, it later appeared) to have kept a chicken heart preparation alive for many years. Given the proper care, he asserted, our cells were immortal [106, 97, 145]. By the 1940s, Carrel's perfusion techniques made it possible for scientists to maintain separate life in the head and the body of a decapitated dog [118, 156].

Carrel not only grew organs outside the body; he used these techniques to preserve organs for transplantation, and also developed the microsurgical techniques that made such transplants possible ([88], p. 733), [58]. By 1920, kidney transplants in animals, skin grafts, corneal transplants, and blood transfusions in people were all a reality; heart and brain transplants were already the subject of serious public speculation ([106], pp. 305, 317; [18], pp. 539, 550; [32]). The first renal transplant between two living patients came in 1954 ([184], p. 50).

As they had since the seventeenth century ([10], pp. 353–361), such organ separations and interchanges seriously complicated the definition of individual life. Many commentators regarded Carrel's work as proving the nineteenth-century claim that the concept of a unified individual was merely an illusion. Embryologist Charles S. Minot wrote, "[I]ndividuality . . . does not exist in nature, except as a rather fantastic notion of the human mind." Thomas Alva Edison agreed. "[W]e are not individuals any more than a great city is an individual. If you cut your finger and it bleeds, you lose cells. They are the individuals" ([74], p. 62).

But for neurologists, these organ separation and transplant experiments strongly implicated the brain as being the locus of the integrative activity that constituted individual life ([125], pp. 27–28). With the development of the electroencephalogram (EEG) in 1929, neurophysiologists first gained the ability to measure a function of the brain directly, without relying on other organs as indicators of the brain's vitality ([184], p. 41). Within a year, George W. Crile had redefined life as the ability to generate spontaneous electric currents [43] and, by 1941, *Time Magazine* reported that the absence of EEG activity was being used as a test for death [241, 223]. EEG studies also sharpened the nineteenth-century conflicts over whether consciousness, or unconscious reflexive integration, was the key to individual life ([106], pp. 313ff.; [80], p. 219; vs. [119]), and over the degree to which each of these activities could be localized within the brain [197, 257].

If transplantation muddied the concept of individual life, this century's remarkable progress in resuscitation made a definition of death seem even more elusive. Three generations of Russian and Soviet scientists, from Bachmetieff in 1910 to Negovskii in the 1960s, produced startling breakthroughs in reviving the seemingly-dead, based on their research on suspended animation. From their discovery that deep cold enabled animals, including humans, to survive an hour or more without vital signs, to the introduction of effective emergency cardio-pulmonary resuscitation, this research both saved lives and complicated the definition of death ([6]; [151], esp. p. 117).

Important suspended animation and resuscitation research took place in the United States and in France as well [7, 18, 19, 108]. In the 1920s and early 1930s, Edward Drinker and colleagues at Harvard built the first practical mechanical respirators, "Iron Lungs," for use in the increasingly severe polio epidemics of the era ([148], p. 8; [162], pp. 324–334). By 1927, European and American physicians studying the problem of industrial electrocution accidents succeeded in using electric stimulation to reverse ventricular fibrillation ([115], p. 6). Such cases grew more common in the late 1940s; and soon, the cardiac defibrillator became the cornerstone of the first hospital intensive care units ([181], p. 233).

Scientists did not shrink from discussing the philosophical implications of these new discoveries. Negovskii asserted that Soviet preeminence in resuscitation research derived from Marxist acceptance of the materialist and mechanistic definition of life implied by true suspended animation. The Russian scientists appropriated as their own Fothergill's 1745 stopped-clock metaphor for the phenomenon [20], ([151], p. 156). On the other hand, Alexis Carrel found in suspended animation, not Marxist materialism, but rather proto-fascist arguments for a dictatorship of doctors. It is profoundly illustrative of the differences between Carrel's time and our own that Carrel, whose discoveries produced so much uncertainty in our definitions of death, never experienced that uncertainty himself. Rather, he became increasingly convinced of the infallibility of science and of scientists. By 1935, he demanded the replacement of democracy with scientocracy; he ended his career as head of a eugenic institute in Nazi-occupied France [35, 58].

The mass media, too, focussed on the conceptual philosophical implications of these new discoveries. In the 1910s it was widely reported that suspended animation "so closely resembl[es] death as to be indistinguishable from it" [108]. Early news accounts of cardiac resuscitation presented it as a form of resurrection [53, 85, 90, 109]. From 1911 to the 1950s, popular articles reported these discoveries under headlines such as "What Is Death?" Many of these publications admitted that doctors lacked a definitive definition of death, and that diagnostic mistakes were still being made [119, 156, 241, 244].[38]

Yet, prior to the late 1960s, such uncertainties were presented so as to emphasize the fascinating mysteries of science, rather than the terrors and dilemmas of the unknown. In past centuries, the gaps in our understanding of death provoked mass panic; but throughout the first half of this century, such mysteries evoked only the "gee whiz" of awe-filled optimism. The unknown dimensions of death seemed to hold, not the threat of premature burial, but the hope of immortality, eternal youth, resurrection, even travel through time.[39]

Despite such media optimism, however, the new discoveries were already starting to cause clinical problems. As early as the 1890s physicians had found occasional brain injury victims whose heart and lung functions could be maintained by continuous artificial respiration, but whose consciousness never returned [47, 66, 105, 172]. With the spread of improved resuscitation devices, such cases became common enough for the International Congress of Anesthesiologists to seek guidance from Pope Pius XII in 1957. His response left it up to doctors to clarify the definition of death. But even if such patients were considered alive, the Pope declared, no "extraordinary" means need be used to maintain them, if they were deemed hopeless cases [173], ([184], pp. 44, 134; [125], pp. 28–29). By 1959, French neurophysiologists had given a name, "coma dépassé," to this syndrome, defined as the loss of all reflexes and of all electrical activity in the brain [147], ([125], p. 29; [180], p. 22).

Also in the 1950s, the legal system first began to grapple with the problem of distinguishing various types of coma from death. Although the English-speaking countries generally had refrained from adopting the elaborate Continental legislation for avoiding premature burial, common law courts for centuries had occasionally been called on to determine if and when a particular individual had died. The area of law known as "survivorship" dealt with such cases, which arose when a person was missing, or when the terms of a will required determining which of two people had died first. Prior to this century, what made these cases controversial was the lack of direct evidence about the actual moment of death. Most of the examples reported involved people lost at sea. In lieu of specific evidence, some courts accepted medical guess-work based on general physiological principles. Other judges relied on formal legal presumptions, such as the ancient rule that a person missing seven years was legally dead. Neither approach was fully satisfactory; and neither had anything to say about clinical medical definitions of death [34], ([5], Vol. 15, p. 2482).

But in the mid-twentieth century, automobile accidents and improved medical care of the comatose combined to help create a new type of survivorship case, where the controversy was due not to a lack of evidence, but to a disagreement over the definition of death. In such early cases as *Thomas v. Anderson* (96 Cal. App. 2d 371, 215 P. 2d 478 (1950)), and *Smith v. Smith* (229 Ark. 579, 317 SW 2d 275 (1958)), courts initially ruled that a comatose patient remained alive so long as heart and lung vital signs were maintained ([184], pp. 46–47). However, the development of organ transplantation complicated these legal issues, beginning as early as 1963. The *Potter* case of that year involved a British physician who had removed a kidney for transplantation from a respirator-maintained brain-damaged donor ([184], p. 51; [11], p. 1950).[40] By the late 1960s, legal definitions of death were in turmoil.

The attempt to resolve a specific clinical problem – how to treat victims of persisting non-responsive brain damage – led to conflation of other issues, such as euthanasia, with the debate over defining death. As the Pope's response of 1957 illustrated, the clinical issues could be handled either by clarifying the definition of death, or by revising the criteria for terminating treatment of the living.

Euthanasia had begun to influence the premature burial debate as early as the 1890s. By the early twentieth century, as euthanasia advocates broadened their concerns to include "quality of life," eugenics, and social costs, instead of just the termination of suffering, the questions of withholding treatment from the living and defining death became increasingly intertwined. Even the famed pioneer advocate for the physically handicapped, Helen Keller, came close to denying that brain-damaged people had the capacity for human life. In her 1915 defense of a doctor who refused to treat a brain-damaged, physically handicapped newborn, she declared, "It is the possibilities of happiness, intelligence and power that give life its sanctity, and they are absent in the case of a poor . . . unthinking creature" [166].

As early as the 1930s, physicians had been forced to use "quality of life" criteria in allocating the limited supply of Iron Lungs available for polio victims. The March of Dimes, a massive fundraising effort inaugurated by President Roosevelt, ended this shortage for polio victims. But physicians soon faced similar resource allocation questions in other chronic respirator cases ([162], pp. 330–334; [125], pp. 18–19).

In short, virtually all of the technical discoveries and social developments that seemed to overwhelm our concept of death in 1967 were already known or clearly predicted between 1910 and the 1950s. Yet, despite continued professional and lay media coverage, nothing like the panic of 1740–1850 occurred in this century, prior to the late 1960s.

Death Since 1967: Heart-Lung, Whole-Brain, or Higher Brain Definitions?

What changed in the late 1960s? The unique cultural significance of the heart and the consequent special meaning of heart transplants surely played a major role in raising public and professional awareness of the problems in defining death. But the revival of public suspicion of medical experts that began in the late 1960s was also extremely important. The nineteenth-century premature burial panic had been ended by a unique period of popular enthusiasm for medical science, and public deference to the judgments of doctors. This era of deference was an almost unprecedented aberration in the history of American medicine. By the end of the 1960s, the medical profession once again faced public criticism on a variety of issues, including the question of defining death.

Initially, public critics revived the ancient fear that doctors would diagnose live people as dead, a result of carelessness, conceptual inadequacies, covert euthanasia, or transplant organ harvesting. *Newsweek*'s 1967 article "When Are You Really Dead?" quoted an unnamed public health official, "I have a horrible vision of ghouls hovering over an accident victim with long knives unsheathed, waiting to take out his organs . . ." [245].

Soon, however, other segments of the public (advocates of euthanasia, the "death-with-dignity" movement, and proponents of health care cost containment) began to criticize physicians for being too conservative in their death diagnoses. Today, for perhaps the first time, public criticism involves both a fear of being wrongly declared dead, and a fear of being wrongly declared alive.

Over the past two decades since the first heart transplants, doctors, lawyers, ethicists, and the public became embroiled in a world-wide debate over redefining death. Though the evolution of this controversy has been studied at great length elsewhere [184–185, 125] ([180], esp. pp. 20–30, 61–71), the key events and issues should be reexamined briefly, in the light of the issue's long previous history.

From the late 1960s to the mid 1970s, most of the efforts to redefine death – including the London Ciba Symposium of 1966, the 1968 World Medical Association meeting in Sydney, Australia, and a variety of state legislative initiatives – developed as specific responses to organ transplantation. They thus inadvertently convinced both the participants and the public that the issue was unprecedented, and derived entirely from new medical technology ([184], pp. 52–72 and passim).

In 1968, a Harvard University committee produced the best known of these early efforts at redefinition. Their report, written by anesthesiologist Henry K. Beecher, offered a set of criteria for diagnosing "irreversible coma." The "Harvard Criteria" emphasized the loss of *all* brain functions: conscious sensation, spontaneous motion, and integrative reflexes. The EEG was also used, but only to supplement various specific tests for each of the specific functional losses ([184], pp. 61–72).

While the Harvard Criteria attempted to measure the termination of both conscious and unconscious brain activity, many other people took the term "brain death" to mean primarily the permanent loss of consciousness. As early as 1966 an article in *Time Magazine* explained brain death as meaning "when the human spirit is gone" [243]. Volition, sensation, and intellect seemed closer to defining the "human spirit" than did such primitive integrative reflexes as yawning, swallowing, and blinking. In 1971, Scottish neurologist J. B. Brierley, writing in the *Lancet*, urged that brain death be defined not as the loss of all brain functions, but as the permanent cessation of "those higher functions of the nervous system that demarcate man from the lower primates . . ." ([184], p. 99).

The 1981 Presidential Commission report, *Defining Death*, rejected this "higher functions" approach, in favor of a "whole-brain" definition that emphasized integrative reflexes. The Commission pointed to lack of agreement on the components of consciousness, the inability of physicians to test reliably for it, the lack of knowledge about which specific brain structures were required for it, and the fact that, unlike whole-brain death, the permanent loss of consciousness was apparently compatible with the indefinite continuation of vitality in other bodily organ systems ([180], esp. pp. 38–40).

Critics of the Commission argued that such limitations were merely temporary shortcomings of our present scientific capabilities and not flaws in the concept itself ([258], pp. 254–256). But others have cogently replied that our subjective states of consciousness are inherently inaccessible to other people. According to this philosophical tradition, there will never be a way to test for the irreversible loss of sensitivity, not because we lack the technology, but because the goal is fundamentally unattainable [116].[41]

The underlying controversy at issue between "whole-brain" and "higher brain" definitions is hardly new. At stake is the question of whether mental consciousness or physiological integration is the basis of individual human life—a question that, as we have seen, has been debated ever since people first distinguished between the death of a person and the death of the body's parts.

The Commission stressed the need to reach a consensus definition in order to replace a series of confusing and contradictory local laws and policies with a uniform, nationally-acceptable approach. They carefully assured that their report would be published with the prior endorsement of the American Medical Association, American Bar Association, and key elements of the "Right to Life" movement ([180], pp. 2, 11; [184], p. 308). In less than five years, most American jurisdictions have adopted their approach.

Yet the apparent success of this consensus-building has not ended the controversy. Efforts to formulate "higher brain" alternatives have continued, especially in the academic community. Doubts about the validity of any brain-based definitions of death have also persisted, particularly outside the medical world, based on both theological and non-theological arguments [199]. While there is no evidence that death tests for whole-brain functions are any more inaccurate than traditional tests for heart-lung functions ([180], p. 82), highly publicized mistakes seemingly have been made in declaring people brain dead ([11], p. 1952); [13], ([184], pp. 58, 73, 310–311; [196], p. 42; [199], p. 272). The initial press reports on brain death heavily emphasized such errors and uncertainties, and the image stuck [54] [245]. And there are still some troubling gaps in our scientific knowledge about whole-brain death. The President's Commission listed eight not-uncommon conditions, from extreme cold to anesthetic overdose, known to cause suspended animation, and thus capable of invalidating a determination of death ([180], pp. 165–166; for others *see* [137], pp. 15, 20). The mechanisms by which such factors operate are still not fully understood. Very recent work even suggests that in at least some vertebrates, adult brains are capable of regenerating lost nerve cells [152]. As a result, popular fears of being wrongly declared dead, even on supposedly "conservative" whole-brain criteria, still can claim some scientific support [11], ([184], p. 58).

Historical Summary and Policy Conclusions

Both the definition and the diagnosis of death have always been uncertain. Yet, particular historical periods, such as the era between 1740 and the 1850s, and the decades since 1967, have produced unusual levels of concern and controversy about these uncertainties. In both periods, new scientific discoveries in resuscitation, suspended animation, and experimental physiology, contributed in complex and previously misunderstood ways to the growth of such anxieties. Historians of the earlier premature burial panic have tended to overlook its scientific origins and to overstate its scientific termination. Accounts of the post-1967 era have grossly inflated the technical factors.

Furthermore, similar physiological and technical discoveries created important scientific uncertainties about death in the century between the 1850s and the 1950s, without producing either medical or public alarm. In each of these eras, a wide variety of social factors, particularly the changing level of public trust in science and deference to doctors, interacted with the technological changes to shape both lay and professional perceptions of death. The premature burial scare of the eighteenth century began among scientists and physicians; by the late nineteenth century, it had become a vehicle for opposition to the scientific professions. The panic petered out early in this century, largely due to increased public faith in doctors. But as that faith and deference waned in the past two decades, controversies over death reemerged.

What can this history tell us about the present? First, since there never was a Golden Age of Hearts and Lungs when defining death was unambiguous and certain, it is probably unrealistic to make the restoration of this mythic consensus the main goal of current policy. Second, the period of social and professional confidence in medical expertise from which we have recently emerged was an historical aberration not likely to recur anytime soon. Lay mistrust of doctors' definitions has been the historical rule rather than the exception, though such mistrust only periodically caused great alarm. Third, death is not simply a timeless and permanently definable term. Its meaning has changed over time, in response to changing technology, social structure, and values. The history of past changes should lead us to expect future changes, any of which might invalidate both the specific diagnostic tests and the basic definitions that we adopt for today's use.

Fourth, and finally, this history reminds us that untestable definitions of death have tended to be profound sources of mischief. From the panic provoked by the alleged lack of tests for death in the eighteenth century, to the introduction of tests for the undefined state of mind called "intelligence" early in our own century, the promulgation of untestably vague definitions of various mental functions simply encouraged the use of poorly conceived available testing expedients that often resulted in serious social harm. By analogy with such precedents, if "higher brain" definitions are not currently capable of being rigorously operationalized, it may be socially harmful and therefore premature to promote them.

University of Michigan
Ann Arbor, Michigan

NOTES

* I have benefitted from unpublished manuscripts graciously shared with me by Rosa Lynn Pinkus, Tom Tomlinson, and Paul Panadero.

[1] Alexander jumbled both Winslow and Bruhier's names ([3], p. 26), and the mistake was repeated by the President's Commission ([180], p. 13). To trace the publishing history, *see* [253–256, 28, 30], [62]. Vol. 14, p. 450.

[2] For similar endorsements of putrefaction, *see* [103], p. 40; [225], p. 791; [158], p. 151.

[3] For doubters even of putrefaction, *see* [124], p. 101; [42], pp. 14–15; [215], p. 81; [3], p. 26.

[4] The theme also appealed to nineteenth-century artists; *see* the monumental painting "Mort Apparente," by the Belgian artist Antoine Joseph Wiertz ([115], p. 31).

[5] Bruhier's first vernacular edition of Winslow's pioneering work on the signs of death incorporated Réaumur's pioneering work on artificial respiration, in 1742 ([28], pp. 344–356). For the attribution of this section to Réaumur, *see* [186]. Cooper's first English

translation of Bruhier added the full text of Tossach's first English-language artificial respiration report, originally published only a year earlier ([254], p. 83; [232]). In his brief introduction to the second English edition, Cooper specifically cited artificial respiration as a primary source of the new uncertainty about death ([256], p. v).

[6] The early history of humane societies may be found in [2, 14, 81, 128, 230].

[7] For typical examples, *see* [124], p. 95; [158], pp. 152–154; [42], pp. 9, 14; [225], pp. 804–805.

[8] No full history of this subject is available, but *see* [203], p. 364; [124], p. 122; [212], pp. 28–29; [169], p. 178; [115], p. 201.

[9] [44], p. 28 – freezing; [209], p. 214 – tetanus, freezing, head trauma, apoplexy; [222], p. 503 – freezing; [78], vol. 1, p. 467 – head trauma, psychosomatic causes, freezing, exhaustion; [103], pp. 31–32 – apoplexy, syncope, opiates, alcohol; [212], pp. 19–23; [67], pp. 6–8 – asphyxia, syncope, apoplexy, hysteria, hypochondriasm, but *not* trance or drugs; [122], p. 168 – asphyxia, hysteria, hypochondria, syncope, hemorrhage, tetanus, apoplexy, epilepsy, and ecstasy.

[10] For similar views, *see* [225], p. 804; [78], vol. 1, p. 380; [17], p. 549; [18], p. 538.

[11] *See* [126], p. 45; [227], p. 219; [170], p. 37; [254], p. 18. In 1606, Shakespeare had King Lear use a looking glass and a feather to search for signs of life in Cordelia; Act V, scene iii, lines 261–266.

[12] The original work on auscultation death tests was [23]. *See* [140] for an early English endorsement, and [123], p. 74 for an early attack. For use of the pulse, *see* [126], p. 45; and for a defense of the pulse as more reliable than the heartbeat, as late as 1889, *see* [192], p. 110.

[13] Credit is usually, but probably erroneously, given instead to Carl Caspar Crève ([2], p. 21; [201]).

[14] For example: Fothergill ([230], p. 44); Cullen ([44], p. 4); Hawes ([103], p. 30); and Kite ([124], pp. 107–108). For others, *see* [92], p. 104; [42], p. 1.

[15] For Charles Kite, muscular irritability alone was the vital principle ([124], pp. 107–108). New Yorker David Hosack preferred John Brown's term, "nervous excitability" ([128], p. 431).

[16] From different premises a century earlier, the German philosopher Gottfried von Leibniz reached similar conclusions about the federated nature of the individual and the gradualness of death. [96], Vol. 2, pp. 18–19; [149], p. 60.

[17] For more on Bichat's very influential theories of life and death, *see* [68], Vol. 1, p. 576; [222], p. 497; [2], p. 21; [62].

[18] Philip's goal was to reassure that death was, by definition, painless; *see also* [74], pp. 54–57; [167], p. 48. The life of the individual was also sometimes identified with the "coenesthesis" – the "sense-of-self" by which we both perceive and integrate our internal sensations [204].

[19] One early-nineteenth-century physician hoped that better criteria for determining death would permit him to stop treatment sooner. It is not clear whether he was concerned about wasting his time and effort, or about violating the Hippocratic injunction against treating the terminally ill ([212], p. 25).

[20] One of the better known early devices was "Bateson's belfry" ([222], p. 493). For others, *see* [122], p. 165; [40], pp. 105–106; [137], p. 19; [180], p. 14.

[21] Many stated quite explicitly that they thought being burned alive far preferable to

premature burial; see [254], pp. 144, 196; [215], pp. 72, 146; [10], p. 362; [227], Chapter 11; [137], p. 19.

[22] For more on the variety of responses within the Jewish community, see [192], pp. 100, 116; [227], pp. 179–181, 406; [249], p. 23.

[23] To trace the growth of confidence in such tests, see [187], p. 147; [82], pp. 405, 407; [177], p. 92.

[24] To trace the growth of confidence in these signs, see [187], p. 148; [99], p. 97; [75]; [115], p. 151; [26], Chapter 3.

[25] For a chorus of similar attacks on suspended animation, see [187], p. 146; [99], p. 88; [211]; [82], p. 402; [177], p. 88; [123], p. 75.

[26] Brown-Séquard had a similar-sounding case as early as 1879 ([227], p. 427).

[27] The parallel has been extensively studied in [222]. For various nineteenth-century observations on the similarity, see [177], p. 90; [26], p. 44; [227], p. 155; [25], pp. 31–33.

[28] Richardson correctly anticipated that the major application of his discoveries would be in food storage technology. His research on hypothermic suspended animation eventually did play a role in the commercial development of frozen food. For biographical data, see [61–62].

[29] One group of patients often linked with those suffering death-like states was the so-called "fasting girls," who allegedly survived long periods without ingesting or excreting. For these and other supposedly entranced young women, see [25], p. 58; [86]; [227], pp. 45, 52; [107], p. 560; [26], p. 27.

[30] For other similar lengthy lists of causes, see [123], p. 74; [26], p. 44; [161], pp. 19, 22; [177], pp. 89–91.

[31] For similar arguments from British and American doctors, see [177], pp. 95–96; [140]; [170], p. 37; [107], p. 564; [157].

[32] For other examples of SPPMB press coverage, see [249], pp. 22–27; [31]; [237]; [137], p. 18.

[33] They were unambiguous in their insistence on this point; see [38], p. 9; [101], pp. i, 104, 127; [249], p. 21; [227], pp. 11, 142; [237, 250, 95].

[34] Richardson was close to several SPPMB members through his involvement in sanitarian public health reform, and his interest in the scientific study of the occult. But he avoided their rabid anti-materialism, and carefully strove to be on both sides of the premature burial issue [221, 61, 62].

[35] Another who combined an interest in burial inspection with the entire range of antimedical reforms was Charles M. Higgins of Brooklyn, President of the Medical Liberty League ([120], p. 467). For the anti-professionalism, spritualism, and reform activities of the Marquis d'Ourches, see [211].

[36] For similarly gruesome preventive measures, see [101], pp. 116–128; [249], pp. 14, 16; [227], p. 272, ch. 21; [250].

[37] In both cases Wilder also traced his concern to the terror evoked by fictional accounts read in childhood; see [248], p. 35; [249], p. 15.

[38] In a mordant essay on mortality in 1919, H. L. Mencken observed that Carrel's work made it "hard to say just when he is fully dead" ([141], p. 237).

[39] For a small sampling of the genre, see [143, 106, 145, 50, 156, 241, 178, 39, 87, 72]. There were occasional equivalents earlier; see Ben Franklin's comments in American Museum (September 1790), 8, 133–134.

[40] For a similar case from Sweden *see* [216].
[41] For a nineteenth century version of this argument, *see* [101], p. 104. For more on Jonas' position, *see* [184], p. 72.

BIBLIOGRAPHY

1. Ackerknecht, E. H.: 1955, *A Short History of Medicine*, Ronald Press, New York.
2. Ackerknecht, E. H.: 1968, 'Death in the History of Medicine', *Bulletin of the History of Medicine* **42**, 19–23.
3. Alexander, M.: 1980, '"The Rigid Embrace of the Narrow House": Premature Burial & The Signs of Death', *Hastings Center Report* **10** (June), 25–31.
4. Altmann, A.: 1973, *Moses Mendelssohn*, University of Alabama Press, University, Alabama.
5. *American Digest, Century Edition*: 1897–1904, West Publishing Company, St. Paul, Minnesota.
6. 'Anabiosis: A State Betwixt Life and Death': 1914, *Review of Reviews* **50** (August), 240–242.
7. 'Anabiosis – Life in Death': 1914, *Literary Digest* **49** (August 22), 304–305.
8. 'Apparent Death': 1869, *Every Saturday* **8** (July 24), 104–107.
9. Ariès, P.: 1975, 'The Reversal of Death: Changes in Attitudes toward Death in Western Societies', in D. E. Stannard (ed.), *Death in America*, University of Pennsylvania Press, Philadelphia, pp. 134–158.
10. Ariès, P.: 1981, *The Hour of Our Death*, Knopf, New York.
11. Arnold, J. D., Zimmerman, T. F., and Martin, D. C.: 1968, 'Public Attitudes and the Diagnosis of Death', *Journal of the American Medical Association* **206**, 1949–1954.
12. Augustine, St.: 1972, *The City of God Against the Pagans*, Loeb Classical Library, Harvard University Press, Cambridge.
13. 'Back From the Dead': 1967, *Newsweek* **70** (November 13), 99.
14. Baker, A. B.: 1971, 'Artificial Respiration, The History of an Idea', *Medical History* **15**, 337–351.
15. Baldwin, J. F.: 1896, 'Premature Burial', *Scientific American* **75** (October 24), 315.
16. Barker, C. F.: 1897, 'On Real and Apparent Death', *Clinique* (Chicago) **18**, 232–238.
17. Beatty, T. E.: 1845, 'Dead (Persons Found)', in J. Forbes (ed.) *Cyclopaedia of Practical Medicine*, Lea & Blanchard, Philadelphia, 1, pp. 549–552.
18. Becquerel, P.: 1915, 'Latent Life: Its Nature and Its Relations to Certain Theories of Contemporary Biology', *Annual Report of the Smithsonian Institution for 1914*, Government Printing Office, Washington, D.C., pp. 537–551.
19. 'Becquerel's Revolutionary Theory of the Significance of Suspended Life': 1916, *Current Opinion* **60** (March), 182–183.
20. 'Between Life and Death': 1913, *Scientific American* **109**, 362.
21. Bichat, M.-F.-X.: 1809, *Physiological Researches Upon Life and Death*, Smith & Maxwell, Philadelphia.
22. Billings, J. S.: 1972, 'The Registration of Vital Statistics (1883)', in G. H. Brieger

(ed.), *Medical America in the Nineteenth Century*, Johns Hopkins Press, Baltimore, pp. 301–310.

23. Bouchut, E.: 1849, *Traité des Signes de la Mort et des Moyens de Prévenir les Enterrements Prématurés*, Bailliére, Paris.
24. Bouchut, E.: 1883, *Traité des Signes de la Mort et des Moyens de Prévenir les Inhumations Prématurées*, 3rd ed. Bailliére, Paris.
25. Braid, J.: 1850, *Observations on Trance; Or, Human Hibernation*, John Churchill, London.
26. Brouardel, P.-C.-H.: 1897, *Death and Sudden Death*, William Wood, New York.
27. Brown, E. M.: 1983, 'Neurology and Spiritualism in the 1870s', *Bulletin of the History of Medicine* 57, 563–577.
28. Bruhier d' Ablaincourt, J.-J. (trans. and add.): 1742, *Dissertation Sur L'Incertitude des Signes de la Mort, Et L'Abus des Enterremens et Embaumemens Précipités*, J. B. Winslow (auth.), Morel, Prault, Prault, and Simon, Paris.
29. Bruhier d' Ablaincourt, J.-J.: 1745, *Memoire Sur La Nécessité D'Un Réglement Génerál Au Sujet des Enterremens, Et Embaumemens*, Simon, Paris.
30. Bruhier d' Ablaincourt, J.-J.: 1749, *Dissertation Sur L'Incertitude des Signes de la Mort, Et L'Abus des Enterremens et Embaumemens Précipités*, 2nd ed., 2 vols., De Bure l'Aîné, Paris.
31. 'Burying Alive': 1895, *The Spectator* 75, 332.
32. 'Can the Brain Be Made to Function After Death?': 1913, *American Review of Reviews* 48 (December), 732–733.
33. Capron, A. M.: 1978, 'The Development of Law on Human Death', in J. Korein (ed.), *Brain Death: Interrelated Medical and Social Issues*, Annals of the New York Academy of Sciences 315 (November 17), pp. 45–61.
34. Cardozo, B. N.: 1894, 'Identity and Survivorship', in A. M. Hamilton and L. Godkin (eds.), *A System of Legal Medicine*, E. B. Treat, New York 1, pp. 213–242.
35. 'Carrel's Man': 1935, *Time* 26 (September 16), 40–43.
36. Carrington, H.: 1910, 'Death: Its Phenomena', *Annals of Psychical Science* 9, 255–267.
37. Cassedy, J. H.: 1965, 'The Registration Area and American Vital Statistics; Development of a Health Research Resource, 1885–1915', *Bulletin of the History of Medicine* 39, 221–231.
38. Chapin, H. G.: 1898–99, 'The Proposed Law to Prevent Premature Burial', *Medico-Legal Journal* (New York) 16, 1–9.
39. Clarke, A. C.: 1958, 'Gone Today, Here Tomorrow: Suspended Animation', *Holiday* 23 (February), 28.
40. Coffin, M. M.: 1976, *Death in Early America: The Hist... and Folklore of Customs and Superstitions*, Thomas Nelson, Nashville.
41. Cohen, A.: 1984, 'Descartes, Consciousness and Depersonalization: Viewing the History of Philosophy From a Straussian Perspective', *Journal of Medicine and Philosophy* 9:1, 7–27.
42. Colhoun, S.: 1823, *An Essay on Suspended Animation*, Edward Parker, Philadelphia.
43. Crile, G. W.; Telkes, M.; and Rowland, A. F.: 1930, 'The Physical Nature of Death', *Scientific American* 143 (July), 30–32.

44. Cullen, W.: 1784, *A Letter to Lord Cathcart . . . Concerning the Recovery of Persons Drowned and Seemingly Dead*, C. Elliott, Edinburgh.

45. Curran, W. J.: 1983, 'An Historical Perspective on the Law of Personality and Status with Special Regard to the Human Fetus and the Rights of Women', *Milbank Memorial Fund Quarterly/Health and Society* **61**:1, 58–75.

46. Curry, J.: 1815, *Observations on Apparent Death*, 2nd ed., E. Cox, London.

47. Cushing, H.: 1902, 'Some Experimental and Clinical Observations Concerning States of Increased Intracranial Tension', *American Journal of the Medical Sciences* **124**, 377–391.

48. *Cyclopaedia of American Biography*: 1915, new ed., Press Association, New York.

49. Dagi, T. F.: 1985, 'Hearts and Souls: The Historical Evolution of the Imperative to Resuscitate', unpublished paper, American Association for the History of Medicine, Chapel Hill, North Carolina.

50. Dallas, H. A.: 1927, 'What Is Death', *Living Age* **332** (February 15), 354–359.

51. Darnton, R.: 1968, *Mesmerism and the End of the Enlightenment in France*, Harvard University Press, Cambridge.

52. 'The Dead-Alive': 1858, *Lancet*, 2:561.

53. '"Dead," He Revives Only to Die Again': 1953, *New York Times* (February 16), p. 12, col. 6.

54. 'Death and Suspended Animation': 1968, *Science News* **94** (August 24), 177–178.

55. 'Death: Its Criteria and Consequences': 1834, *Medical Magazine* (Boston) **2**, 369–388, 441–456.

56. *Dictionary of American Biography*: 1959 and supps., Scribner, New York.

57. *Dictionary of American Medical Biography*: 1928, H. A. Kelly and W. L. Burrage (eds.), Appleton, New York.

58. *Dictionary of American Medical Biography*: 1984, M. Kaufman, S. Galishoff, and T. Savitt (eds.), Greenwood Press, Westport, Connecticut.

59. *Dictionary of the History of Ideas*: 1968, P. P. Wiener (ed.), Scribner, New York.

60. *Dictionary of the History of Science*: 1981, W. F. Bynum, E. J. Browne, and R. Porter (eds.), Princeton University Press, Princeton, New Jersey.

61. *Dictionary of National Biography*: 1921 and supps., Oxford University Press, Oxford.

62. *Dictionary of Scientific Biography*: 1981, C. C. Gillispie (ed.), Scribner, New York.

63. *Dictionnaire de Biographie Française*: 1932–, Librairie Letouzey et Ané, Paris.

64. *Dictionnaire des Biographies*: 1958, P. Grimal (ed.), Presses Universitaires de France, Paris.

65. Ducachet, H. W.: 1822, 'On the Signs of Death, and the Manner of distinguishing real from apparent Death', *American Medical Recorder* (Philadelphia) **5**, 34–53.

66. Duckworth, Sir D.: 1898, 'Some Cases of Cerebral Disease in Which the Function of Respiration Entirely Ceases For Some Hours Before That of the Circulation', *Edinburgh Medical Journal* N. S. **3**:2, 145–152.

67. Dunglison, [R.]: 1827, *Syllabus of the Lectures on Medical Jurisprudence and on The Treatment of Poisoning & Suspended Animation, Delivered in the University of Virginia*, University of Virginia, [Charlottesville].

68. Dunglison, R.: 1833: *A New Dictionary of Medical Science*, 2 vols., Charles Bowen, Boston.

69. Ellenberger, H.: 1970, *Discovery of the Unconscious*, Basic, New York.
70. Engelhardt, H. T., Jr.: 1975, 'John Hughlings Jackson and the Mind-Body Relation', *Bulletin of the History of Medicine* **49**, 137–151.
71. Erhardt, C. L., and Berlin, J. E. (eds.): 1974, *Morbidity and Mortality in the United States*, Harvard University Press, Cambridge.
72. Ettinger, R. C. W.: 1966, 'Can Deep Freeze Conquer Death', *Ebony* **21** (January), 60–61.
73. Eyler, J. M.: 1979, *Victorian Social Medicine: The Ideas and Methods of William Farr*, Johns Hopkins University Press, Baltimore.
74. Farrell, J. J.: 1980, *Inventing the American Way of Death, 1830–1920*, Temple University Press, Philadelphia.
75. 'Fear of Being Buried Alive: Infallible Signs of Death': 1920, *Scientific American Monthly* **1** (May), 396–397.
76. Ferriar, J.: 1798, 'Of the Treatment of the Dying', in *Medical Histories and Reflections*, Vol. 3, Cadell and Davis, London, pp. 189–208.
77. Flew, A.: 1964, *Body, Mind, and Death: Readings*, Macmillan, New York.
78. Forbes, J. *et al.*: 1845, *Cyclopaedia of Practical Medicine*, 4 vols., Lea & Blanchard, Philadelphia.
79. Fothergill, J.: 1980, *Observations on the Recovery of a Man Dead in Appearance by Distending the Lungs with Air* (1745), Richard J. Hoffman, San Francisco.
80. Fox, W. U.: 1984, *Dandy of Hopkins*, Williams & Wilkins, Baltimore.
81. France, E. M.: 1975, 'Some Eighteenth Century Authorities on the Resuscitation of the Apparently Drowned', *Anaesthesia* **30**, 530–538.
82. Fraser, W.: 1880–81, 'Distinctions Between Real and Apparent Death', *Popular Science Monthly* **18**, 401–408.
83. French, R. D.: 1975, *Antivivisection and Medical Science in Victorian Society*, Princeton University Press, Princeton.
84. French, S.: 1975, 'The Cemetery as a Cultural Institution: The Establishment of Mt. Auburn and the "Rural Cemetery" Movement', in D. E. Stannard (ed.), *Death in America*, University of Pennsylvania Press, Philadelphia, pp. 69–91.
85. 'From Beyond the Styx: English Gardener Tells of Experience While Heart Stopped': 1935, *Literary Digest* **119** (February 23), 26.
86. Gairdner, W. T.: 1883–84, 'Case of Lethargic Stupor, or "Trance"', *Lancet* **2**, 1078–1080, and **1**, 56–58.
87. Galston, A. W.: 1965, 'Death Isn't Necessary', *Science Digest* **58** (July), 80–85.
88. Garrison, F. H.: 1929, *An Introduction to the History of Medicine*, 4th ed., W. B. Saunders, Philadelphia.
89. Garrison, F. H.: 1969, *Garrison's History of Neurology*, rev. by L. C. McHenry, Jr., Charles C. Thomas, Springfield, Illinois.
90. 'Girl "Dies," Then Lives 22 Hours': 1953, *New York Times* (February 18), p. 56, col. 3.
91. Glantz, L. H.: 1983, 'The Role of Personhood in Treatment Decisions Made by Courts', *Milbank Memorial Fund Quarterly/Health and Society* **61**:1, 76–100.
92. Goodwyn, E.: 1788, *The Connection of Life With Respiration*, T. Spilsbury for T. Johnson, London.
93. Gould, S. J.: 1981, *The Mismeasure of Man*, W. W. Norton, New York.

94. Grier, J.: 1937, *A History of Pharmacy*, Pharmaceutical Press, London.
95. 'The Growing Practice of Premature Burial': 1907, *Current Literature* **43** (December), 680–682.
96. Hall, T. S.: 1969, *Ideas of Life and Matter: Studies in the History of General Physiology 600 B.C. to A.D. 1900*, 2 vols., University of Chicago Press, Chicago.
97. Hallowell, C.: 1985, 'Charles Lindbergh's Artificial Heart', *American Heritage of Invention & Technology* **1** (Fall), 58–62.
98. Hannah, J. E.: 1967, 'The Signs of Death: Historical Review', *North Carolina Medical Journal* **28**, 457–460.
99. Harris, F. A.: 1894, 'Death in its Medico-Legal Aspects' in A. M. Hamilton and L. Godkin (eds.), *A System of Legal Medicine*, E. B. Treat, New York, 1, 57–137.
100. Hartmann, F.: 1895, *Buried Alive: An Examination into the Occult Causes of Apparent Death, Trance and Catalepsy*, Occult Publishing Company, Boston.
101. Hartmann, F.: 1896, *Premature Burial*, Swan Sonnenschein, London.
102. Hauerwas, S.: 1978, 'Religious Concepts of Brain Death and Associated Problems', in J. Korein (ed.), *Brain Death: Interrelated Medical and Social Issues, Annals of the New York Academy of Sciences* **315** (November 17), 329–337.
103. Hawes, W.: 1780, *An Address to the Public on Premature Death and Premature Interment*, for the Author, London.
104. Hawes, W.: 1782, *An Address to the King and Parliament of Great Britain, on the Important Subject of Preserving the Lives of its Inhabitants. . . . Hints for Improving the Art of Restoring Suspended Animation*, J. Dodsley, London.
105. 'Heart-Beats After Death': 1931, *Literary Digest* **110** (August 15), 28.
106. Hendrick, B. J.: 1913, 'On the Trail of Immortality', *McClures's* **40** (January), 304–317.
107. Herold, J.: 1899, 'Signs and Tests of Death', *New Orleans Medical and Surgical Journal* **51**, 549–566, 609–623, 669–687.
108. Hirshberg, L. K.: 1913, 'Between Life and Death', *Lippincott* **92**, 651–652.
109. Hoehling, A. A.: 1955, 'She Was Dead for 50 Minutes', *Reader's Digest* **66** (May), 49–52.
110. Hunter, J.: 1776, 'Proposals for the Recovery of Persons Apparently Drowned', *Philosophical Transactions of the Royal Society of London* **46**, 412–425.
111. Huston, K. G.: 1976, *Resuscitation: An Historical Perspective*, Wood Library-Museum, Park Ridge, Illinois.
112. [Interview with Alexis Carrel]: 1935, *Newsweek* **6** (December 21), 41.
113. [Interview with Alexis Carrel]: 1935, *Time* **26** (December 23), 24.
114. Jackson, C. O.: 1977, *Passing: The Vision of Death in America*, Greenwood Press, Westport, Connecticut.
115. Jellinek, S.: 1947, *Dying, Apparent-Death and Resuscitation*, Williams & Wilkins, Baltimore.
116. Jonas, H.: 1974, 'Against the Stream: Comments on the Definition and Re-definition of Death', in *Philosophical Essays*, Prentice-Hall, Englewood Cliffs, New Jersey, pp. 132–140.
117. Josat, [A.]: 1854, *De la Mort et de Ses Caractères*, Germer Bailliére, Paris.
118. Kaempffert, W.: 1953, 'Physicians Are Faced With More Uncertainty As to the Fact

of Clinical Death', *New York Times* (February 22), Sect. IV, p. 9, col. 6.

119. Kaempffert, W.: 1953, 'What Is Clinical Death?', *Science Digest* **34** (August), 76–77.

120. Kaufman, M.: 1967, 'The American Anti-Vaccinationists and Their Arguments'. *Bulletin of the History of Medicine* **41**, 463–478.

121. Keller, E. F.: 1985, *Reflections on Gender and Science*, Yale University Press, New Haven.

122. Kennedy, J. G.: 1977, 'Poe and Magazine Writing on Premature Burial', *Studies in the American Renaissance*, pp. 165–178.

123. Kesteven, W. B.: 1855, 'The Signs of Death', *British and Foreign Medico-Chirurgical Review* **15**, 71–77.

124. Kite, C.: 1788, *An Essay on the Recovery of the Apparently Dead*, C. Dilly, London.

125. Korein, J.: 1978, 'The Problem of Brain Death: Development and History', in J. Korein (ed.), *Brain Death: Interrelated Medical and Social Issues, Annals of the New York Academy of Sciences*, **315** (November 17), 19–38.

126. [Lancisi, G. M.]: 1971, *De Subitaneis Mortibus (On Sudden Deaths)* (1707), P. D. White and A. V. Boursy (trans.), St. John's University Press, New York.

127. Laycock, T.: 1860, *Mind and Brain, or the Correlations of Consciousness and Organization*, 2 vols., Sutherland and Knox, Edinburgh.

128. Lee, R. V.: 1972, 'Cardio-pulmonary Resuscitation in the Eighteenth Century', *Journal of the History of Medicine and Allied Sciences* **27**, 418–33.

129. 'Live Hearts in Dead Bodies': 1928, *Literary Digest* **98** (July 7), 23.

130. 'Liver Kept Alive Outside a Body': 1953, *New York Times* (February 22), Sect. IV, p. 9, col. 6.

131. Lloyd, J. U.: 1910, *The Eclectic Alkaloids*, Bulletin of the Lloyd Library No. 12, J. U. and C. G. Lloyd, Cincinnati.

132. Louis, [A.]: 1752, *Lettres Sur la Certitude des Signes de la Mort, où l'on rassure les Citoyens de la crainte d'être enterrés vivans; avec des observations & des expériences sur les noyés*, Michel Lambert, Paris.

133. McCandless, P.: 1981, 'Liberty and Lunacy: The Victorians and Wrongful Confinement', in A. Scull (ed.), *Madhouses, Mad-Doctors, and Madmen: The Social History of Psychiatry in the Victorian Era*, University of Pennsylvania Press, Philadelphia, pp. 339–362.

134. McLellan, I.: 1981, 'Nineteenth-Century Resuscitation Apparatus', *Anaesthesia* **36**, 307–311.

135. McManners, J.: 1981, *Death and the Enlightenment: Changing Attitudes to Death among Christians and Unbelievers in Eighteenth-Century France*, Oxford University Press, New York.

136. Madden, T. M.: 1897, 'On Morbid Somnolence and Death-Trance', *Medical Magazine* (London) **6**, 857–862, 922–929.

137. Marshall, T. K.: 1967, 'Premature Burial', *Medico-Legal Journal* **35**, 14–24.

138. Massachusetts General Court: 1904, *Journal of the House of Representatives*.

139. Massachusetts General Court: 1904, *Journal of the Senate*.

140. "M.D.": 1847, 'The Prevention of Premature Burials', *Littell's Living Age* **13**, 357–358.

141. Mencken, H. L.: 1955, 'Exeunt Omnes', from the *Smart Set* (1919), in A. Cooke

(ed.), *The Vintage Mencken*, Vintage Books, New York, pp. 233–240.

142. Mendelsohn, E.: 1964, *Heat and Life: The Development of the Theory of Animal Heat*, Harvard University Press, Cambridge.

143. Metchnikoff, E., and Williams, H. S.: 1912, 'Why Not Live Forever?', *Cosmopolitan* **53** (September), 436–446.

144. Mettler, L. H.: 1890–91, 'The Import, From the Medico-Legal Standpoint Of Distinguishing Between Somatic and Molecular Death', *Medico-Legal Journal* (New York) **8**, 172–179.

145. Middleton, J.: 1914, 'Flesh That is Immortal: Dr. Alexis Carrel's Experiments With Tissues of a Chicken', *World's Work* **28** (October), 590–593.

146. Miller, J.: 1978, *The Body in Question*, Random House, New York.

147. Mollaret, P., and Goulon, M.: 1959, 'Le Coma Dépassé', *Revue Neurologique*, **101**:1, 3–15.

148. Mörch, E. T.: 1985, 'History of Mechanical Ventilation', in R. R. Kirby, R. A. Smith, and D. A. Desautels (eds.), *Mechanical Ventilation*, Churchill, Livingstone, New York, pp. 1–58.

149. Morison, R. S.: 1971, 'Definitions of Death and the Quality of Life; Issues Raised by Organ Transplantation and Hemodialysis,' in *Medicine and Society: Contemporary Medical Problems in Historical Perspective*, American Philosophical Society Library, Philadelphia, pp. 51–66.

150. Morris, R. J.: 1976, *Cholera, 1832*, Holmes & Meier, New York.

151. Negovskii, V. A.: 1962, *Resuscitation and Artificial Hypothermia*, Consultants Bureau, New York.

152. 'New Neurons Form in Adulthood': 1984, *Science* **224**, 1325–1326.

153. New York State Legislature: 1871, *Assembly Journal*, 94th sess., Albany.

154. New York State Legislature: 1871, *Senate Journal*, 94th sess., Albany.

155. New York State Legislature: 1898, *Assembly Journal*, 121st sess., Albany.

156. Newman, B. M.: 1940, 'What Is Death?', *Scientific American* **162** (June), 336–337.

157. 'Only One Sure Sign of Death': 1930, *Literary Digest* **106** (August 30), 27.

158. Orfila, [M. J. B.]: 1818, *A Popular Treatise on the Remedies to be Employed in Cases of Poisoning and Apparent Death; Including the Means of . . . Distinguishing Real from Apparent Death*, William Phillips, London.

159. Orfila, M. J. B.: 1819, *Directions for the Treatment of Persons Who Have Taken Poison, and Those in a State of Apparent Death . . . also of Distinguishing Real from Apparent Death*, Nathaniel G. Maxwell, Baltimore.

160. *Oxford English Dictionary, Compact Edition*: 1971, Oxford University Press, Oxford.

161. Parry, L. A.: 1914, 'The Possibility of Premature Interment', *Medical Magazine* (London) **23**, 11–23.

162. Paul, J. R.: 1971, *A History of Poliomyelitis*, Yale University Press, New Haven.

163. Payne, L. M.: 1971, 'The Moment of Death: An Historical Commentary', *Medical and Biological Illustration* **21**, 45–51.

164. Pearson, J. W.: 1965, *Historical and Experimental Approaches to Modern Resuscitation*, Charles C. Thomas, Springfield, Illinois.

165. Peebles, J. M.: 1913, *Vaccination a Curse and a Menace to Personal Liberty*, 10th ed., Peebles, Los Angeles.

166. Pernick, M. S.: 1985, '"The Black Stork": An American Movie and the Debate Over Withholding Medical Treatment from Deformed Newborns, 1915–1927', unpublished paper presented at American Association for the History of Medicine, Chapel Hill, North Carolina, and forthcoming in *Bulletin of the History of Medicine*.

167. Pernick, M. S.: 1985, *A Calculus of Suffering: Pain, Professionalism, and Anesthesia in Nineteenth-Century America*, Columbia University Press, New York.

168. Péron-Autret, Dr.: 1979, *Les Enterrés Vivants: Histoires des Morts Vivants ou les Incertitudes des Signes de la Mort*, Balland, [Poitiers].

169. Philip, A. P. W.: 1834, 'On the Nature of Death', *Philosophical Transactions of the Royal Society* **1**, 167–198.

170. 'The Physical Phenomena of Death': 1850, *Eclectic Magazine* (January), 23–52.

171. Pinkus, R. L.: 1984, 'Families, Brain Death, and Traditional Medical Excellence', *Journal of Neurosurgery* **60**, 1192–1194.

172. Pinkus, R. L.: 1985, 'Brain Death', *Journal of Neurosurgery* **62**, 160–161.

173. Pius XII: 1958, 'The Prolongation of Life', *The Pope Speaks* **4** (November 24), 393–398.

174. Poe, E. A.: 1975, 'The Premature Burial' (1844), in *The Complete Tales and Poems*, Vintage Books, New York, pp. 258–268.

175. Poe, E. A.: 1975, 'The Facts in the Case of M. Valdemar', in *The Complete Tales and Poems*, Vintage Books, New York, pp. 96–103.

176. Poole, W. F.: 1958, *Poole's Index to Periodical Literature 1802–1881*, Peter Smith, Gloucester, Massachusetts.

177. Porter, G. L.: 1882, 'Recognition of Death', *Proceedings of the Connecticut Medical Society*, N.S. 2:3, 87–96.

178. 'Predicts Rip Van Winkles May Be Real in 50 Years': 1956, *Science Newsletter* **69** (February 25), 121.

179. 'Premature Burial': 1904, *Medical Press and Circular* (London) **77** (April 27), 458.

180. President's Commission for the Study of Ethical Problems in Medicine and Biomedical and Behavioral Research: 1981, *Defining Death: Medical, Legal, and Ethical Issues in the Determination of Death*, Government Printing Office, Washington, D.C.

181. President's Commission for the Study of Ethical Problems in Medicine and Biomedical and Behavioral Research: 1983, *Deciding to Forego Life-Sustaining Treatment: Ethical, Medical, and Legal Issues in Treatment Decisions*, Government Printing Office, Washington, D.C..

182. 'The Problem of Burial Alive': 1905, *Current Literature* **39** (July), 66–67.

183. Proctor, R. A.: 1879, 'Suspended Animation', *Contemporary Review* **36**, 501–520.

184. Rado, L. A.: 1979, 'Communication, Social Organization and the Redefinition of Death', unpublished Ph.D. dissertation, University of Pennsylvania.

185. Rado, L. A.: 1981, 'Death Redefined: Social and Cultural Influences on Legislation', *Journal of Communication* **31**, 41–47.

186. Réaumur, R. -A. -F. de: 1752, *Avis Pour Donner du Secours à Ceux que l'on Croit Noyés* (1740), reprinted in [A.] Lôuis, *Lettres Sur la Certitude des Signes de la Mort*, Michel Lambert, Paris, pp. 250–260.

187. Reese, J. J.: 1879, 'The Signs and Phenomena of Death', *Hospital Gazette* (New York) **6**, 145–148, 161–164.

188. Reid, J.: 1841, *The Philosophy of Death: General Medical and Statistical Treatise on*

the Nature and Cause of Human Mortality, S. Highley, London.

189. Reiser, S. J.: 1978, *Medicine and the Reign of Technology*, Cambridge University Press, Cambridge.

190. Rendell-Baker, L.: 1981, 'Nineteenth-Century Resuscitation Apparatus', *Anaesthesia* **36**, 1058–1059.

191. Richardson, B. W.: 1879, 'Suspended Animation', *Living Age* **142**, 99–103.

192. Richardson, B. W.: 1888–89, 'The Absolute Signs and Proofs of Death', *Proceedings of the Medical Society of London* **12**, 100–117.

193. [Richardson, B. W.]: 1889, 'The Electric Test of Death', *Asclepiad* **6**, 248–249.

194. [Richardson, B. W.]: 1893, 'The Diaphanous Test of Death', *Asclepiad* **10**, 162–163.

195. Robbins, R. H.: 1970, 'Signs of Death in Middle English', *Mediaeval Studies* **32**, 282–298.

196. Roelofs, R.: 1978, 'Some Preliminary Remarks on Brain Death', in J. Korein (ed.), *Brain Death: Interrelated Medical and Social Issues, Annals of the New York Academy of Sciences* **315** (November 17), pp. 39–44.

197. Rosenfield, I.: 1985, 'The New Brain', *New York Review of Books* **32** (March 14), 34–38.

198. Rosenkrantz, B. G.: 1972, *Public Health and the State: Changing Views in Massachusetts, 1842–1936*, Harvard University Press, Cambridge.

199. Rosner, F. and Bleich, J. D. (eds.): 1979, *Jewish Bioethics*, Hebrew Publishing Company, Brooklyn.

200. Roth, N.: 1980, 'Electroresuscitation and the Occult', *Medical Instrumentation* **14**, 120–121.

201. Roth, N.: 1980, 'Life-and-Death Decisions: Electrodiagnosis and Premature Burial', *Medical Instrumentation* **14**, 322.

202. Rothstein, W. G.: 1972, *American Physicians in the Nineteenth Century*, Johns Hopkins University Press, Baltimore.

203. Schechter, D. C.: 1971, 'Early Experience With Resuscitation by Means of Electricity', *Surgery* **69**, 360–372.

204. Schiller, F.: 1984, 'Coenesthesis', *Bulletin of the History of Medicine* **58**, 496–515.

205. See, W.: 1880, 'The Extreme Rarity of Premature Burials', *Popular Science Monthly* **17** (October), 526–530.

206. Selby, R.: 1985, 'The Medical Determination of Death', in R. H. Wilkins and S. S. Rengachary (eds.), *Neurosurgery*, McGraw-Hill, New York, 3, pp. 2585–2597.

207. Sherrington, C. S.: 1906, *The Integrative Action of the Nervous System*, Scribner, New York.

208. Shortt, S. E. D.: 1984, 'Physicians and Psychics: Anglo-American Medical Response to Spiritualism 1870–1890', *Journal of the History of Medicine and Allied Sciences* **39**, 339–355.

209. Shrock, N. M.: 1835, 'On the Signs that Distinguish Real from Apparent Death', *Transylvania Journal of Medicine* **8**, 210–220.

210. Shryock, R. H.: 1947, *The Development of Modern Medicine: An Interpretation of the Social and Scientific Factors Involved*, Knopf, New York.

211. 'The Signs of Death. Report on the Prizes Founded by the Marquis d'Ourches': 1874, *London Medical Record* **2**, 205–207, 221–223.

212. Smith, J. G.: 1821, *Principles of Forensic Medicine*, Underwood, London.

213. Smith, J. G.: 1827, *Principles of Forensic Medicine*, 3rd ed., Underwood, London.

214. Snart, J.: 1817, *Thesaurus of Horror; or, The Charnel-House Explored!!*, Sherwood, Neely, and Jones. London.
215. Snart, J.: 1824, *An Historical Inquiry Concerning Apparent Death and Premature Interment*, Sherwood, Neely, and Jones, London.
216. Snider, A. J.: 1967, 'When is a Person Dead?', *Science Digest* **62** (October), 70–71.
217. Spencer, T. D.: 1881, 'The Phenomena of Death', *Popular Science Monthly* **19**, 394–399.
218. Spicker, S. F. (ed.): 1970, *The Philosophy of the Body*, Quadrangle, Chicago.
219. Stannard, D. E. (ed.): 1975, *Death in America*, University of Pennsylvania Press, Philadelphia.
220. Starr, P.: 1982, *The Social Transformation of American Medicine*, Basic, New York.
221. Stevenson, L.: 1955, 'Science Down the Drain', *Bulletin of the History of Medicine* **29**, 1–26.
222. Stevenson, L.: 1975, 'Suspended Animation', *Bulletin of the History of Medicine* **49**, 482–511.
223. Sugar, O., and Gerard, R. W.: 1938, 'Anoxia and Brain Potentials', *Journal of Neurophysiology* **1**, 558–572.
224. Sutton, G.: 1984, 'The Physical and Chemical Path to Vitalism: Xavier Bichat's Physiological Researches on Life and Death', *Bulletin of the History of Medicine* **58**, 53–71.
225. Symonds, J. A.: 1836, 'Death', in R. B. Todd (ed.), *Cyclopaedia of Anatomy and Physiology*, Longman, Brown, Green, Longmans, & Roberts, London, 1, pp. 791–808.
226. Tebb, W. and Vollum, E. P.: 1896, *Premature Burial and How It May be Prevented, With Special Reference to Trance, Catalepsy, and Other Forms of Suspended Animation*, S. Sonnenschein, London.
227. Tebb, W., and Vollum, E. P.: 1905, *Premature Burial and How It May Be Prevented, With Special Reference to Trance, Catalepsy, and Other Forms of Suspended Animation*, 2nd ed., W. R. Hadwen (ed.), Swan Sonnenschein, London.
228. 'Thesaurus of Horror . . .': 1819, *Blackwood's Magazine* **5**:27, 334–337.
229. Thomas, L.: 1974, *The Lives of a Cell*, Bantam, New York.
230. Thompson, E. H.: 1963, 'The Role of Physicians in the Humane Societies of the Eighteenth Century', *Bulletin of the History of Medicine* **37**, 43–51.
231. Todd, R. B.: 1845, 'Pseudomorbid Appearances', in J. Forbes *et al.*, *Cyclopaedia of Practical Medicine*, Lea & Blanchard, Philadelphia, 3, pp. 723–733.
232. Tossach, W.: 1744, 'A Man, Dead in Appearance Recovered by Distending the Lungs With Air', *Medical Essays and Observations* (Edinburgh) **5**:2, 605.
233. Turner, J.: 1980, *Reckoning with the Beast: Animals, Pain and Humanity in the Victorian Mind*, Johns Hopkins University Press, Baltimore.
234. Twain, M.: 1957, 'A Dying Man's Confession' (1883), in *The Complete Short Stories*, Bantam, New York, pp. 225–240.
235. Twain, M.: 1957, 'The Invalid's Story' (1882), in *The Complete Short Stories*, Bantam, New York, pp. 187–192.
236. Twain, M.: 1957, 'The Story of the Old Ram' (1872), in *The Complete Short Stories*, Bantam, New York, pp. 77–81.
237. "T. W.": 1895, 'Burying Alive', *The Spectator* **75**, 399.
238. Walsh, D.: 1898, *Premature Burial: Fact or Fiction?*, William Wood, New York.

239. *Webster's Biographical Dictionary*: 1963, G. & C. Merriam, Springfield, Massachusetts.
240. 'What is Death?': 1896, *The Spectator* **77**, 933–934.
241. 'What is Death?': 1941, *Time* **37** (June 2), 62.
242. 'What is Death?': 1968, *Scientific American* **219** (September), 85.
243. 'What is Life? When is Death?': 1966, *Time* **87** (May 27), 78.
244. 'What is Meant by Death?': 1911, *Review of Reviews* **43**, 623–624.
245. 'When Are You Really Dead?': 1967, *Newsweek* **70** (December 18), 87.
246. *Who Was Who in America*: 1942 and supps., Marquis Who's Who, Chicago.
247. Wiebe, R. H.: 1967, *The Search for Order; 1877–1920*, Hill & Wang, New York.
248. Wilder, A.: 1871, 'Annual Address of the President of the Eclectic Medical Society', *New York State Assembly Documents*, No. 84, pp. 30–46.
249. Wilder, A.: 1895, *Perils of Premature Burial: A Public Address Delivered Before the Members of the Legislature, at the Capitol, Albany, N.Y., January 25, 1871*, E. W. Allen, London.
250. Williamson, J. R.: 1896, 'Premature Burial', *Scientific American* **74** (May 9), 294.
251. Wilson, L. G.: 1959, 'Erasistratus, Galen and the *Pneuma*', *Bulletin of the History of Medicine* **33**, 293–314.
252. Wilson, L. G.: 1960, 'The Transformation of Ancient Concepts of Respiration in the Seventeenth Century', *Isis* **51**, 161–172.
253. Winslow, J. B.: 1740, *An Mortis Incertae Signa Minus Incerta a Chirurgicis, Quam ab Aliis Experimentis?*, Quillau, Paris.
254. [Winslow, J. B.]: 1746, *The Uncertainty of the Signs of Death and Danger of Precipitate Interment*, M. Cooper, London.
255. [Winslow, J. B.]: 1748, *The Uncertainty of the Signs of Death and Danger of Precipitate Interment*, George Faulkner, Dublin.
256. [Winslow, J. B.]: 1751, *The Uncertainty of the Signs of Death and Danger of Precipitate Interment*, 2nd ed., M. Cooper, London.
257. Young, R. M.: 1970, *Mind, Brain and Adaptation in the Nineteenth Century*, Oxford University Press, Oxford.
258. Youngner, S. J., and Bartlett, E. T.: 1983, 'Human Death and High Technology: The Failure of the Whole-Brain Formulations', *Annals of Internal Medicine* **99**, 252–258.

ROLAND PUCCETTI

DOES ANYONE SURVIVE NEOCORTICAL DEATH?

A person is dead when an irreversible cessation of all that person's brain functions has occurred. The cessation of brain functions can be determined by the prolonged absence of spontaneous cardiac and respiratory functions.

Law Reform Commission of Canada
Working Paper 23 (1979, pp. 58–59)

An individual with irreversible cessation of all functions of the entire brain, including the brain stem, is dead.

Defining Death, President's
Commission ([15], p. 162)

One day on a very cold morning, I started the engine of my car without opening the garage door, then retreated to the kitchen to let it warm up. At that moment there was a long distance phone call from my publisher, and I spent a half hour haggling over the terms of a contract. Upon returning to the garage I found there, alongside the car, my dog Fido. Of course I immediately opened the automatic door and carried him outside into the fresh air. When I saw he wasn't breathing I pressed rhythmically on his chest cavity, and sure enough his heart and lungs started to respond. But still he did not wake up, so I took him to the vet's on the way to the university. The vet said Fido appeared to be comatose as a result of carbonmonoxide poisoning, and advised that I prepare myself for the worst. He said it seemed the dog would not recover consciousness because the top of the brain was destroyed, although the brain stem must be intact, since Fido could breathe unaided. What should I do?

Now suppose I told you that I did the following. I took Fido home, made a special bed for him, shaved his body and fitted him with diapers, learned to feed him intravenously and nasogastrically, and arranged to have him turned in his bed often so he would not develop pressure sores. Evenings I would watch TV alongside Fido, stroking his warm body and listening to his breathing, though never again did he go for a walk with me, fetch the newspaper, bark, or do anything dogs normally

75

Richard M. Zaner (ed.), Death: Beyond Whole-Brain Criteria, pp. 75–90.
© 1988 *by Kluwer Academic Publishers.*

do.[1] Would you think I am a rational person? Or worse, suppose I asked the vet to put Fido down for me and he replied that he couldn't do that to a comatose canine capable of spontaneous breathing, *because the law forbade it*. Would you think this a rational society?

If your answers to those questions are, as I think they would be, resoundingly negative, then presumably the main reason for your negative response is simply that my ministrations would do *no good* for Fido, who has permanently lost consciousness even if he is nonapneic; for the same reason, *no harm* can be done to him by stopping his breathing. In other words, a permanently unconscious dog is, for all practical purposes, a dead dog anyway.

Now change the above apocryphal story in just one detail. Instead of it being Fido on the garage floor, it is my neighbor's infant son, little Bobby. I rush him to a children's hospital, but with the same result. Now could what was agreed to be irrational behavior in Fido's case become rational in Bobby's case? That a human life is much more valuable and worth preserving at all costs is completely irrelevant, since we agreed before that the main consideration was simply that one can do neither good nor harm to an irreversibly comatose being, and that should hold true independently of his species identity.

Yet, and this is what I find amazing, a great number of otherwise very able people in the legal and health professions think just the opposite. But before we examine their arguments, let us get one point absolutely straight. No matter what your calling, any position you adopt on this issue is going to be a *philosophical* stand. For the question we are addressing here is quintessentially philosophical: namely, *what constitutes* (in this world anyway, to avoid begging religious questions) *personal death*? This does not mean a poll of professional philosophers will settle the matter, for by that standard the Earth must have been flat in the Middle Ages, before it became spherical again in the Renaissance. Nevertheless, you cannot offer argument as to when a person may be safely considered dead without doing philosophy. The question, of course, is whether you are doing philosophy well or badly. I begin with an example of what I consider bad philosophizing on this subject.

I. WALTON ON CEREBRAL VS. ENCEPHALIC DEATH

In his recent book [17], Douglas N. Walton asks why proponents of the cerebral or neocortical criterion of brain death are not even more selective.

The reply (that it is the cerebrum which mediates cognitive activity in the brain) is still not entirely satisfactory. However, it seems equally plausible to say that mental activity of the higher cognitive sort takes place essentially in the cerebral cortex, the thin membraneous (*sic*) substance that forms a mantle over the cerebrum. Why include the lobes of the cerebrum under the cortex if lower parts of the midbrain or the cerebellum and brain stem are excluded? . . . [The cerebral death advocate] might argue that it is safer to include the whole cerebrum, because there is a possibility of indeterminacy or error. But then, if tutiorism (chancing error to be on the safe side) is brought in, why not be even safer and take into account the whole-brain ([18], p. 50)?

To which one can counter: if it is better to err on the side of safety, why not wait until complete somatic death (including, e.g., cartilage cells in the knee) occurs, thereby foregoing organ transplantation? The reason cerebral death includes the lobes of the cerebrum beneath the cortical surface is that these are composed of association fibers (white matter) that interconnect neocortical motor and sensory areas (grey matter); if these are selectively destroyed (e.g., in asphyxiation), the result is a shattered self with isolated conscious functions [5]. If, on the other hand, the neocortical surface is itself selectively destroyed, which as we shall see is the case with apallic syndrome, that is sufficient to obliterate all conscious functions.

The case for excluding other subcortical structures mentioned by Walton in passing has nothing at all to do with tutiorism anyway. The midbrain is of course just the top of the brain stem, where the superior and inferior colliculi trigger orientating reflexes related, respectively, to sources of visual and auditory stimuli: such reflexive responses do not require conscious mediation, as we all know from finding ourselves turned towards an abrupt movement in the peripheral visual field, or in the direction of a sudden sound, before such stimuli register in consciousness. If a brain structure does its job unconsciously, then there is no reason to think its integrity in a comatose patient is evidence of residual conscious functions. Similarly with the cerebellum, which preorchestrates complex bodily movements, and under therapeutic electrode stimulation does not yield clear sensations [3]. The cerebellum probably also stores learned subroutines of behavior, like swimming or typing: precisely the kinds of things you do better when not concentrating on them. Why then is it necessary for the cerebellum to be dead in order to have a dead person on your hands?

Yet Walton seems to think that everything above the level of C_1 directly contributes to a conscious mental life. His attitude seems to be that if something moves, it must or at least might have a life of its own, with its own kind of feeling. For example, of the pupillary reflex mediated by the lower brain stem he says:

The pupillary reflex could, for all we know, indicate some presence of feeling or sensation even if the higher cognitive faculties are absent. Even if we cannot resolve the issue with the precision we would like and, indeed, just because of that, we should be on the safe side . . . Following my tutiorist line of argument, it is clear that we cannot rule out the possibility that brain stem reflexes could indicate some form of sensation or feeling, even if higher mental activity is not present ([18], p. 69).

The statement fairly reeks of superstition. As we all know, when the doctor flashes his penlight on the eye, we do not feel the pupil contract, then expand again when he turns the light off. If not, then why in the world does Walton suppose that a deeply comatose patient feels anything in the same testing situation? The whole point of evolving reflexes like this, especially in large brained animals that do little peripheral but lots of central information processing, is to shunt quick-response mechanisms away from the cerebrum so that the animal can make appropriate initial responses to stimuli *before* registering them consciously. If one could keep an excised human eye alive *in vitro* and provoke the pupillary reflex, the way slices of rat hippocampus have been stimulated to threshold for neuronal excitation, would Walton argue that the isolated *eye* might feel something as its pupil contracts?[2]

Apparently he would, since in his view the entire encephalon must be safely dead to justify a finding of personal death. The spinal cord, he agrees, presents no obstacle to a determination of death, because of the relative sparsity of its neurons and .its accessibility to tactile stimuli exclusively. I quote:

The upshot is that brain death should include the whole-brain but nothing more. I would add that specially if we emphasize the element of reflective selfconsciousness or awareness . . . the tactile stimuli accessible from the spinal cord need not be thought significant if, in the absence of a brain, there is no possibility of awareness of these stimuli. The presence of a tactile reflex by itself need not indicate mental activity or consciousness ([18], p. 75).

But what is sauce for the goose is sauce for the gander. Why does a pinprick in the foot or hand, provoking a withdrawal reflex in that limb because of residual electrical activity in the spinal arc nerve pathways, count as a tactile sensation in the case of a cerebrally dead patient, but not in the case of an encephalically dead one? For consider: there is no such thing as a non-localized tactile sensation; to the question, "Where did the accused touch you on your person?" there always is and must be a fairly specific answer. Even a statement like "I felt the spring sun warm my body all over, like a gentle kiss," is in a sense localized: it refers to all the skin surface on one side of the body, that facing the sun. Now what brain mechanism localizes sensations of touch? Surely it is the somatosensory strip posterior to the Rolandic fissure in both cerebral hemispheres (Brodmann's area 3), which is clearly gone in the cerebrally dead patient. No localization, no tactile sensation: thus Walton's unconcern for spinal cord-mediated nerve impulses from the skin surface is logically extendable to the entire subcortical architecture of the brain.

Walton nevertheless expresses deep puzzlement over this matter, and in fact at times comes perilously close to committing the Fallacy of Ad Ignorantium. For example, wondering whether perception or awareness might not occur in the absence of higher cognitive functions, he writes:

It is hard to know how to define exactly what is meant by "higher cognitive faculties," so it is hard to be sure that we mean by this phrase something that definitely excludes all types of perception that might persist in deeper parts of the brain ([18], pp. 75–76).

But surely ignorance of whether a statement is true does not imply that we know it is false; the admitted complexity of the human brain does not carry with it the imputation that we cannot say *some* things about it with confidence. One thing I feel reasonably confident in stating is that sensations are not experienced without recruitment of populations of neurons in the grey matter on the cerebral cortical surface. And it is easy to see why this is so: the phylogenetic novelty of *neo*cortex is due to brain expansion in primates beginning about 50 million years ago to accommodate increasing intelligence, for where else could new cell layers appear but on the outer surface of the brain [9]? That being the case, sensation migrated there as well, and although deeper structures certainly contribute complexly to the sentient input, this is not transduced as sensation until, at a minimum, some 10^4 neurons are provoked to discharge on the surface of at least one cerebral hemisphere at the same time [16]. It is also plain why the contribution of subcortical mechanisms to this input does not itself implicate conscious perception. If it did, we would have sensations in *seriatum*: a baseball leaving the pitcher's hand would be seen as arriving by the hitter several times in succession as neural impulses course from retina to optic chiasm to geniculate body through the optic radiation to primary visual cortex in the occipital lobe. From an evolutionary viewpoint, that would be a recipe for disaster.

Walton makes the claim that, to be on the safe side, "we should presume that the whole-brain is required to produce mental activity" ([18], p. 74). Normally, of course, that is true, but from this it would hardly follow that, when the ultimate neuronal destination of neural input is no longer there, or is dead, sensations still occur. We can liken this proposal to having an express elevator that whisks passengers from the ground floor to executive offices on the top floor. If a nuclear strike blows off the top floor, it would be unreasonable to suppose that you could nevertheless conduct your business at still-standing intermediate floors. What Walton is doing is confusing the normally necessary contribution of subcortical mechanisms to sensation with the sufficient condition of neocortical functions. In the case of the primary visual system in man this is indisputable: destruction of Brodmann's area 17 alone, say by shrapnel wounds, brings permanent total blindness [7]; whereas a peripherally blind person with intact visual cortex can be induced to experience visual sensations by direct electrode stimulation of that grey matter [2].

Probably the sensation we worry most about in irreversibly comatose patients is *pain*. Now Walton may know that depth electrode stimulation intended to relieve intractable "central" or "thalamic" pain (where the pain is not localized at all), indicates that there are discrete "pain centers" in the thalami [14, 17], and these are certainly subcortical structures that could survive neocortical death. Would this fact not tend to show that such patients might nevertheless experience pain, in line with Walton's tutioristic cautions?

Such a suggestion would be, I think, doubly wrong. It is wrong, first, because it incorporates an excessively *homuncular* view of the relation between brain structures and conscious experience. The "little men in the thalami getting pain messages" picture is absurd: it is *we* who get the pains, not those structures. And it is wrong, secondly, because if the firing of thalamic pain centers by itself gave rise to pain experience, then it surely follows that if we could excise this tissue, keep it alive *in vitro*, stimulate it electrically to threshold for discharge and record the neurons' discharging, we would have to say that there is pain going on *in the vat*! Anyone who would believe this is beyond reason.

II. PEOPLE AS BRAIN STEMS

I want now to consider a specific case of neocortical death without brain stem death, in order to show the utter futility of acting on the stand Walton defends. There are many cases like this one, and in some somatic survival subsequent to cerebral destruction far surpasses the present case.[3] Nevertheless, the case history here is remarkably complete. I give it *in extenso* so that I cannot be accused of glossing over some crucial fact [8].

> *Case 8.* The patient (Th. Sv.) was a female who had been born in 1936. In July 1960, at the age of 24, she suffered severe eclampsia during pregnancy with serial epileptic attacks, followed by deep coma and transient respiratory and circulatory failure. In the acute phase, Babinski signs were present bilaterally and there was a transitory absence of pupillary, corneal and spinal reflexes. A left-sided carotid angiogram showed a slow passage of contrast medium and signs of brain edema. An EEG taken during the acute phase did not reveal any electrical activity. The EEG remained isoelectric for the rest of the survival time (seventeen years). After the first three to four months the patient's state became stable with complete absence of all higher functions.
>
> Examination ten years after the initial anoxic episode showed the patient lying supine, motionless, and with closed eyes. Respiration was spontaneous, regular and slow with a tracheal cannula. The pulse was regular. The systolic blood pressure was 75–100 mm. Hg. Severe flexion contractures had developed in all extremities. Stimulation with acoustic signals, touch or pain gave rise to primitive arousal reactions including eye-opening, rhythmic movement of the extremities, chewing and swallowing, and withdrawal reflexes. The corneal reflex was present on the left side. When testing was done on the right side, transient horizontal nystagmus movements were elicited. Pupillary reflexes were present and normal on both sides. On passive movements of the head, typical vestibulo-ocular reflexes were elicited. The spinal reflexes were symmetrical and hyperactive. Patellar clonus was present bilaterally. Divergent strabismus was found when the eyes were opened (by the examiner). Measurement

of the regional cerebral blood flow on the left side (ten years after the initial anoxic episode) showed a very low mean hemisphere flow of 9 ml/100g/min. The distribution of the flow was also abnormal, high values being found over the brain stem. The patient's condition remained essentially unchanged for seven more years and she died seventeen years after the anoxic episode after repeated periods of pulmonary edema.

Autopsy showed a highly atrophic brain weighing only 315 grams. The hemispheres were especially atrophied and they were in general transformed into thin-walled yellow-brown bags. The brain stem and cerebellum were sclerotic and shrunken. On the basal aspect some smaller parts of pre-served cortex could be seen, mainly in the region of the unci. Microscopically the cerebral cortex was almost totally de-stroyed with some remnants of a thin gliotic layer and under-neath a microcystic spongy tissue with microphages containing iron pigment. The basal ganglia were severely destroyed, whereas less advanced destruction was found in the subfron-tal basal cortex, the subcallosal gyrus, the unci, the thalamus and hypothalamus, and in the subicular and entorhinal areas. In the cerebellum the Purkinji cells had almost completely disappeared and were replaced by glial cells. The granular layer was partly destroyed. The cerebellar white matter was partly demyelinated. In the brain stem some neurons had disappeared and a diffuse gliosis was found. Several cranial nuclei remained spared. The long sensory and motor tracts were completely demyelinated and gliotic, whereas trans-verse pontine tracts remained well myelinated ([8], pp. 196–198).

This clinical picture, confirmed by the autopsy findings, is known in the literature as "neocortical death without brain stem death," or more recently and appropriately, as "the apallic syndrome," for its character-istic feature is precisely destruction of the paleum, that cortical mantle of grey matter covering the surface of the cerebrum or telencephalon. As it happens, neurons composing the paleum are the most vulnerable to oxygen deprivation during transient cardiac arrest or, as in the above case, asphyxiation. Whereas in encephalic or whole-brain death, there-fore including the brain stem that monitors respiration (which in turn provokes cardiac activity), the patient can be sustained on a ventilator for only up to a week in adults and two weeks in children before cardiac standstill, the apallic patient breathes spontaneously and demonstrates cephalic reflexes (also brain stem mediated), so that if fed nasogastrically or intravenously and kept free from infection, he or she can sustain somatic life for years or even decades after losing the top of the brain.

I said "somatic life," for without a paleum the basis for a conscious and hence a personal life in this world is gone forever. But then what are we doing supplying intensive care to apallic syndrome patients? For surely the quality of life in a patient like Th.Sv. during all those seventeen years differed not one jot from the quality of life of someone buried underground for seventeen years. It was zero. Permanent unconsciousness is permanent unconsciousness whether the condition is associated with a body that lives by virtue of being able to breathe spontaneously or not. To deny this is to elevate spontaneous breathing to a principle of human life, something I find incredible. Yet this is actually the emerging consensus of the medico-legal community in North America, as indicated by my earlier quotations from the Reform Commission's Report in Canada for 1979, and the U.S. President's Commission's Report of 1981. Are there any good arguments to support this astonishing attitude?

The President's Commission goes out of its way to justify regarding a whole-brain dead person sustained on a respirator as a dead person whose organs are therefore freely transplantable, whereas cerebrally dead persons are still to be considered alive and requiring heroic maintenance procedures. Listen to this passage:

While the respirator and its associated medical techniques do substitute for the functions of the intercostal muscles and the diaphragm, which without neuronal stimulation from the brain cannot function spontaneously, they cannot replace the myriad functions of the brain stem or of the rest of the brain. The startling contrast between bodies lacking *all* brain functions and patients with intact brain stems (despite severe neocortical damage) manifests this. The former lie with fixed pupils, motionless except for the chest movements produced by their respirators. The latter can not only breathe, metabolize, maintain temperature and blood pressure, and so forth, *on their own* but also sigh, yawn, track light with the eyes, and react to pain or reflex stimulation ([15], p. 35).

One is tempted to cry out: *So what*? I can not only breathe, metabolize, maintain temperature and blood pressure, *on my own*, but also sigh, yawn, and react to reflex stimulation (such as the patellar reflex) when in a deep, dreamless sleep. Admittedly I cannot track light with my eyes, or react to pinprick in the foot, without awakening, but that is because, with an intact reticular formation and intact primary visual system, the light I am tracking causes visual sensations, and with an intact somatosensory strip pinprick in the foot wakes me up because I feel it as such. The apallic syndrome patient *can* do both these things without awakening, of course, because those neocortical structures are permanently missing. In fact, he or she cannot do *anything* consciously anymore, because there is no one home in that head to wake up.[4]

This consideration seems to make no difference to members of the President's Commission, who then go on to say:

It is not easy to discern precisely what it is about patients in this latter group that makes them alive while those in the other category are not. It is in part that in the case of the first category (i.e., absence of all brain functions) when the mask created by the artificial medical support is stripped away what remains is not an integrated organism but "merely a group of artificially maintained subsystems." Sometimes, of course, an artificial substitute can forge the link that restores the organism as a whole to unified functioning. Heart or kidney transplants, kidney dialysis, or an iron lung used to replace physically-impaired breathing ability in a polio victim, for example, restore the integrated functioning of the organism as they replace the failed function of a part. Contrast such situations, however, with the hypothetical of a decapitated body treated so as to prevent the outpouring of blood and to generate respiration: continuation of bodily functions in that case would not have restored the requisites of human life ([15], pp. 35–36).

But what *are* the requisites of human life according to the President's Commission? Evidently they are the ability to breathe spontaneously, demonstrate cephalic reflexes, regulate body temperature, metabolism, and blood pressure, etc.: in other words, *to perform unaided the janitorial functions of the Central Nervous System, which requires no conscious direction or reflection whatsoever*. Thus by the standards of the Commission, a hypothetical decapitated human body treated so as to prevent the outpouring of blood, but with brain stem left intact so that these janitorial functions are performed unaided, qualifies as a live person with all the requisites of human life, whereas if someone lops off the stem and substitutes a mechanical respirator in its place, the patient becomes a dead person thereby. Again, what is the difference in quality of life, or even prospects for the quality of life, in the two cases? And again the answer seems undeniable: none whatever.

To sum up: either human life is rooted in brain stem function or it is rooted in the capacity for personal experience. If the former, then all vertebrate species are on an equal footing and what counts in medical ethics is just long-term organic functioning independently of our capacity to intervene. If the latter, then there is no ethically relevant difference in the status of encephalically and cerebrally dead people: they have both lost the neocortical basis of an ongoing personal life. The only surprising fact to come out of the apallic syndrome is, or should be, that corpses are really of two kinds: the vast majority that cannot breathe unaided, and a small minority that nevertheless can do this. Apneic or nonapneic, a corpse is still a corpse.

III. THE DIAGNOSTIC PROBLEM

In a still more recent writing Walton [19] returns to defense of the whole-brain criterion of death. He begins by restating his earlier scepticism about our ability to avoid negative error in diagnosis of the apallic syndrome. He says:

However, physicians have not yet developed or tested proven, certainly safe criteria for the apallic syndrome or other so-called "vegetative states" more highly localized in dysfunction than whole-brain death. For the present, tutioristic reasoning dictates cleaving to criteria for whole-brain death ([19], p. 270).

I do not know what Walton understands by "certainly safe" criteria for diagnosis of apallic syndrome; since the practice of medicine is an empirical science, there is always room for error. However, an accumulation of diagnostic results like those given in the case of Th. Sv. seems strongly conclusive. If one can safely exclude the possibility of hypothermia or intoxication with a C.N.S. depressive, which drastically lowers oxygen requirements of the neocortex, a repeated finding of diminished cerebral blood flow, to less than 20% of normal, is itself powerful evidence of pallial destruction. In fact Lassen *et al.* [10] reported firm correlations between increases in blood flow to specific regions of the cortex of patients in the waking, conscious state when problem solving, and also a significant overall increase in blood flow to the cerebrum as a whole of about 10 per cent during such activity. It therefore seems exceedingly unlikely that any kind of conscious experience is going on in a brain with use for less than a fifth of normal blood flow.[5]

However, with the advent of Positron Emission Tomography, all reasonable grounds for doubt can be removed. In PET scanning, the uptake of oxygen and particularly glucose in selective subregions of cerebral cortex can be measured and displayed in color on a video screen: yellow or green for normal metabolic activity, blue or purple for low or no uptake. Although as of a couple years ago the PET scan had apparently not been used to confirm a diagnosis of apallic syndrome [4], nothing could be safer, easier or more certain: the entire surface of the brain would be displayed in blue or purple. While such deep subcortical structures as the brain stem itself are not visualizable using the PET scan [11], the absence of apnea is by itself evidence of the integrity of the stem. So much for Walton's concerns about negative error in diagnosing apallic syndrome.

IV. SEEDS OF DOUBT

Before leaving Walton's contrary stand on this issue, I want to reply to two further points he makes in that more recent defense of the whole-brain criterion. Neither is really very important, but they illustrate the tendency in philosophical debate to sow confusion among your opponents by almost any means at hand.

First, Walton says that even in the face of a repeated isoelectric EEG, where there is not death of the whole-brain there can be *restoration* of cortical activity through reactivation of the brain stem arousal system. The imputation, of course, is that cortical destruction may itself be reversible. Where there is breath there is always hope.[6]

But Walton refers to only one case [6]. And he does not go on to say that the patient, victim of a motorcycle fall, was treated with electrode stimulation for only 19 days following five months' akinetic mutism, gave only partial signs of arousal, and at no time recovered spontaneous motor activity, leading to abandonment of the treatment. If this patient *had* recovered consciousness, then he could not have been correctly diagnosed as a case of post-traumatic apallic syndrome, for the fact is that no one, after age 16 or so, sprouts new central neurons. The failure in this case to secure sustained arousal is indeed confirmation of post-traumatic pallial destruction, and remains, sadly, incurable.

Walton's second fresh point in the debate alludes to findings by Lober (reported in [12]), that some people recovered from infantile hydrocephaly, thus growing up with severely reduced cerebral hemispheres, can nevertheless function well: an example being that of a university student, IQ 126, who gained first class honors in mathematics. This Walton takes to be evidence that the neocortex is neither the sole seat of consciousness nor, perhaps, crucial to the return of conscious functions.

One wants to scream aloud a commonplace of clinical psychopathology: When neural plasticity enters the picture, all bets are off! The neural plasticity of the infant brain allows a lot less than the normal quantity of grey matter to take over a wide range of functions that are usually diffused in greater brain space. This is strikingly and uncontroversially demonstrated in complete hemispherectomy for infantile hemiplegia, where control of the whole body (except for distal finger movements in the arm contralateral to the missing half brain) is found in adulthood [1]. Furthermore, as Epstein has said (quoted in [12]), hydrocephalus is principally a disease of the *white* matter of the brain (the cerebral ventricles, swelled by overproduction of cerebrospinal fluid, disrupt the axons of association fibers around them). It is precisely the *sparing* of nerve cells in the grey matter, even in severe cases of hydrocephalus, that explains the retention of conscious functions and high-performance IQs.

To summarize against Walton on both these points, *for those who have a normal history of neocortical development*, the integrity of the neocortex is essential to the continuance of a mental, and hence a personal, life. It follows from this that pallial destruction is equivalent to personal demise, and this has nothing to do with a residual capacity for spontaneous respiration. Thus both the wholly brain dead and the cerebrally dead patient are dead people, and it is only superstition to make a vital dichotomy between them.

V. ETHICAL CONSIDERATIONS

Many who have followed me so far would nevertheless balk at the problem of disposing of human remains capable of breathing spontaneously. They would say that active intervention to stop the breathing prior to preparation for burial is not only presently illegal (laws can be changed, and already have, to facilitate organ harvesting), but morally murder; and it is often argued that passive (by the non-continuance of treatment) as opposed to active euthanasia is the more humane course.

Both these replies miss the point. You can stab a corpse, but you cannot *kill* it, for it is already dead whether breathing or not. If neocortical death is agreed to constitute personal death, then a firmly diagnosed apallic syndrome patient is in no better or worse situation than the encephalically dead patient sustained on a respirator, whom we all agree is dead though still breathing artificially. Similarly, "euthanasia" means "mercy killing," but one cannot be merciful to a cadaver, for cadavers are beyond pain and indeed all further experience of this world. Indeed, if it were *my* nonapneic remains causing the problem, I want to insist in advance that the breathing be stopped, for by treating my body as if *I* were still alive, hospital personnel would be stripping me of human dignity: it is enough to have others change your diapers in the first years of life.[7]

Someone might suggest that the difference between defenders of the whole-brain criterion and defenders of the neocortical standard is that they envisage different logical subjects: the former taking that to be a still living, spontaneously breathing human *body*, the latter a *mind* now gone from this world. But if this were just a verbal dispute, it ought not to matter much to the disposal problem in a more enlightened age. But in fact it does matter. Those taking the first view would have to say to enquiring relatives and friends, "He's still alive but permanently unconscious, so we're going to let him die of dehydration, starvation, or infection, whichever comes first." Whereas those taking the second view, which is my own, would logically respond in words like these: "She's dead but her body is still breathing, so we're going to stop the breathing and prepare her body for burial."

Needless to say, the latter formulation seems to me less cruel.

Dalhousie University
Nova Scotia, Canada

NOTES

[1] He might, however, occasionally wag his tail. Apparently this is spinal cord mediated, as dogs coming out of anesthesia often wag their tails. I owe this suggestion to John Fentress.

[2] Of course no such contraction would occur, since the pupillary reflex is brain stem mediated; the point is strictly hypothetical.

[3] McWhirter gives the following case: "The longest recorded coma was that undergone by Elaine Esposito (b. Dec. 3, 1934) of Tarpon Springs, Florida. She never stirred after an appendectomy on Aug. 6, 1941, when she was 6, in Chicago, Illinois, and she died Nov. 24, 1978 aged 43 years 357 days, having been in a coma for 37 years 111 days" ([13], p. 37).

[4] Conversely, selective damage to the brain stem alone can produce a similar result, for without input from the brain stem's reticular activating system, untouched neocortical cells cannot be alerted to incoming stimuli and the patient never wakes up. In the apallic syndrome the lines, so to speak, are still up but no one is home; whereas with brain stem lesions sparing cells that monitor respiration, there is someone home but permanently slumbering because the lines are down. In either case a personal life in this world has ended, so the same remarks made in the section of this paper entitled "Ethical Considera-

tions" will apply. It is perhaps confusion of the necessary contribution of the brain stem reticular formation to achieving conscious awareness with conscious awareness itself that motivates superstitious attitudes towards the apallic syndrome.

[5] D. H. Ingvar has recently reported that cerebral blood flow in apallic patients has been measured using Xenon 133 gas inhalation and 254 scintillators that monitor oxygen and glucose uptake in as many square centimeters of superficial cerebral cortex on each side of the head. The video display is uniform: dark blue or purple on the entire screen, indicating little or no uptake of blood, because "these patients have no neocortex to supply with blood" (Ingvar, in a Symposium organized by the Faculty of Medicine, University of Montreal, entitled "Two Hemispheres: One Brain," on May 18, 1984).

[6] An anatomist once suggested to me that there is hope for eventual restoration of cortical activity in experiments where embryonic fetal tissue has been successfully transplanted into homologous lesioned areas of the adult rat brain. While this may be encouraging with regard to treatment of, say, expressive aphasia, I do not see how one could hope to replace an entire neocortex that way; surely the survivor of such an operation would not be the original person.

[7] This point was originally supplied by a former student of mine who has nursed apallic syndrome patients (name withheld).

BIBLIOGRAPHY

1. Basser, L. S.: 1962, 'Hemiplegia of Early Onset and the Faculty of Speech with Special Reference to the Effects of Hemispherectomy', *Brain* **85**, 427–460.
2. Brindley, G. A. and Lewin, W. S.: 1968, 'The Sensations Produced by Electrical Stimulation of the Visual Cortex', *Journal of Physiology* **196**, 479–493.
3. Cooper, I. S. *et al.*: 1974, 'The Effect of Chronic Stimulation of Cerebellar Cortex on Epilepsy in Man', in I. S. Cooper, M. Riklan and R. S. Snider (eds.), *The Cerebellum, Epilepsy, and Behavior*, Plenum Press, New York and London, pp. 119–171.
4. Feindel, W.: 1982, Personal Communication.
5. Geschwind, N. *et al.*: 1968, 'Isolation of the Speech Area', *Neuropsychologia* **6**, 327–340.
6. Hassler, R. *et al.*: 1969, 'Behavioural and EEG Arousal Induced by Stimulation of Unspecific Projection Systems in a Patient with Post-traumatic Apallic Syndrome', *Electroencephalography and Clinical Neurophysiology* **27**, 306–310.
7. Holmes, G.: 1945, 'The Organization of the Visual Cortex in Man', *Proceedings of the Royal Society* (Biology) **132**, 348–361.
8. Ingvar, D. H. *et al.*: 1978, 'Survival After Severe Cerebral Anoxia, with Destruction of the Cerebral Cortex: The Apallic Syndrome', *Annals of the New York Academy of Sciences* **315**, 184–214.
9. Jerison, H. J.: 1973, *Evolution of the Brain and Intelligence*, Academic Press, New York and London.
10. Lassen, N. A. *et al.*: 1978, 'Brain Function and Blood Flow', *Scientific American* **239**, 62–71.
11. LeBlanc, M.: 1983, Personal Communication.
12. Lewin, R.: 1980, 'Is Your Brain Really Necessary?', *Science* **210**, 1232–1234.
13. McWhirter, N. D.: 1984, *Guinness Book of World Records*, Bantam, New York.

14. Melzack, R.: 1973, *The Puzzle of Pain*, Penguin, London.
15. President's Commission for the Study of Ethical Problems in Medicine and Biomedical and Behavioral Research: 1981, *Defining Death: Medical, Legal, and Ethical Issues in the Determination of Death*, U.S. Government Printing Office, Washington, D.C.
16. Puccetti, R.: 1981, 'The Case of Mental Duality: Evidence from Split-brain Data and Other Considerations', *The Behavioral and Brain Sciences* **4**, 92–123.
17. Sem-Jacobsen, C.: 1968, *Depth-Electrographic Stimulation of the Human Brain and Behavior*, Thomas, Springfield, Illinois.
18. Walton, D. N.: 1980, *Brain Death: Ethical Considerations*, Purdue University Press, West Lafayette, Indiana.
19. Walton, D. N.: 1981, 'Epistemology of Brain Death Determination', *Metamedicine* **2**, 259–274.

H. TRISTRAM ENGELHARDT, JR.

REEXAMINING THE DEFINITION OF DEATH AND BECOMING CLEARER ABOUT WHAT IT IS TO BE ALIVE

The papers by Martin Pernick and Roland Puccetti are more complimentary than may at first be apparent. Pernick portrays the interplay, through time, of four major clusters of questions regarding death. In the process, his paper provides a history that leads to the issues examined by Roland Puccetti. These four clusters of questions are: (1) What does it mean to be alive? (2) What does it mean to be embodied? (3) What tests will provide us with an acceptably low number of false-positive tests for being dead? and (4) How does one choose a point of death along a continuum from being alive to having died? These questions in turn depend for their answers on the meaning of being alive. Human life has variously been interpreted as equivalent to: (a) the presence of a soul as an animating and rational entity, (b) the presence of a soul as a vital principle, (c) life as biological integration, and (d) life as the presence of a person.

The history of the definition of death reflects these ambiguities, which are due in part to cultural conflicts among various religious and metaphysical views of life and death. In addition, there is philosophical unclarity regarding the differences between human life as biological integration versus human life as a certain level of sentience or consciousness that is integral to the existence of persons. The first can be seen as a vitalist interpretation of human life and the second as a personalist interpretation of human life. At stake in the contrast between these interpretations is the meaning of being alive as a human individual. The more one concludes that what is uniquely of significance about us as humans is our life as persons, the more the personalist's interpretation becomes the one to embrace.

There has been an evolution in our appreciation of the embodiment of life. The issue of embodiment is tied to the first question: depending on whether one is speaking of life simpliciter or the life of a person, one will be brought to explore a different sense of embodiment. We as a culture came better to understand biological functions as the vitalist's

91

Richard M. Zaner (ed.), Death: Beyond Whole-Brain Criteria, pp. 91–98.
© 1988 by Kluwer Academic Publishers.

interpretation of life became less plausible. One might think in particular of Friedrich Woehler's synthesis of urea in 1828. He showed that life was not necessary for the synthesis of organic compounds. It became clear that what was unique to humans was not some special vitalizing principle, but rather a particular level of sentience or, in fact, self-consciousness. Even here, there has been an evolution. One might consider Pernick's remarks concerning Greek medical understandings of the heart as the seat of life and certain feelings. This past understanding and its images are still influential in our culture today, as the role of hearts on Valentine cards attests. Given our current understandings, a more appropriate metaphor might be a display of the limbic system. Or to take a second example, one might consider the Roman Catholic devotion to the Sacred Heart of Jesus. However, it is the brain, not the heart, that is the direct sponsor of consciousness, moral sentiments, feelings of affection, and the dedication displayed in the religious life.

The history of the last two and a half millennia has been the history of the better appreciation of the embodiment of those mental functions we identify with the life of persons. We have moved from the notion of the embodiment of mental faculties in various parts of the body and from the view that all of soul is in all of the parts of the body (e.g., St. Thomas Aquinas's position that the soul is *tota in toto, tota in qualibet parte*) to the view that the capacities we identify with persons are embodied in the brain. The 19th century saw, in fact, a mediation between the Aristotelian view of different souls representing vegetative, animal, and rational integration on the one hand and the Platonic view of the soul as the directing conscious mind on the other. The 19th century produced an understanding of the mind-brain relationship that saw the consciousness of persons embodied in the sensory-motor integrative capacities of the brain, while recognizing different levels of integration. The unique functions of persons were seen to be sponsored by the brain so that vitalist understandings of biological integration were abandoned by many.

We are led then to the problem of determining when the embodiment of persons has been destroyed. As all actual tests of actual occurrences, there will be risks of false-positive and false-negative determinations. A choice of a particular test will depend on how much one is willing to risk being declared dead prematurely. Of course, such judgments should also take into consideration the costs of false-negative tests of death, which will lead to sustaining the biological life of the bodies of individuals who have died, but which the test cannot show to be dead. For example, I, and I presume Puccetti as well, would hold that Karen Quinlan was dead for years before she was legally declared dead. The continued sustenance of her body was an example of the consequences of a false-negative determination.

Whether one will tolerate such false-negative findings will depend, in part, on whether one holds that one has a test that can declare individuals such as Karen Quinlan dead without exposing us to significant false-positive determinations of being dead. Finally, there is the problem of deciding when death occurs along a continuum of changes. This question will need to be answered in part in terms of how certain one needs to be that the necessary conditions for the life of a person have irrevocably been destroyed. To answer this question, one will need to become clearer about what functions are to be measured and to fashion precise and accurate measurements of those functions whose absence indicates the destruction of the embodiment of a person.

Contemporary discussions of the definition of death have tended to focus on the death of persons. This is not unexpected in that only persons write books on death, philosophies regarding life's meaning, or religious tracts on the immortality of the soul. Persons as self-conscious entities have an unavoidable centrality in philosophical considerations, which centrality is not a mere cultural quirk. Persons are integral to the very notion of the moral enterprise. There would be no morality without persons, and only in terms of persons can morality be articulated. It is for this reason that the destruction of the brain is essential to modern definitions of death.

Scientific advances have made it clear that the brain is the sponsor of consciousness and other organs can be transplanted without disturbing the identity of persons. But this insight leads to the question of whether one is alive, just because a part of the brain is still functioning. This point can be put in plain language in terms of a good news-bad news story. Consider being informed by a neurologist that one is suffering from an untreatable, irreversible neurological disease that will destroy one's whole-brain. The good news is that through the wonders of modern technology and a substantial research endowment, one can be kept alive for a normal life expectancy. If one concludes that such life would not only be of no significance or value to oneself, but that in addition, one would not *be there*, then one has understood and accepted the meaning of whole-brain death. One can imagine, then, that a second opinion shows that the bad news is not really that bad. One's entire brain will not be destroyed. The entire brain stem will be able to be salvaged so that one can live a full life expectancy breathing on one's own, unassisted by mechanical ventilation. If one still comes to the same conclusion, namely, that not only would such life be of no use to oneself, but that in addition one would not *be* there living in the body, then one will have taken the first step to a higher-brain-centers-oriented concept of death. One will be recognizing that one's interest in a definition of death focuses primarily on those parts of the brain that are the sponsors of consciousness. One will have acknowledged that the mere presence of one's brain stem will not be sufficient for one's continued existence. One will have recognized that one means by the life of persons not just mere biological integration, but at least minimal sentience, if not minimal self-consciousness. One will have taken a firm step to accepting the distinction between being alive as a person versus mere human biological life.[1]

This distinction is philosophical. In principle, one could have made the distinction in ancient Greece. Still, the distinction would not have had its contemporary character without the development of modern neurophysiology and modern concepts of the cerebral localization of mental functions. Such distinctions were at hand by the end of the 19th century and the beginning of the 20th century. Yet, the whole-brain definition of death did not develop till the latter part of the 20th century. Its development at that time was tied in part to the economic and public policy considerations that forced us to become clearer about what we mean by being alive in the world. When it became possible to sustain brain dead but otherwise alive human bodies, it became necessary to determine whether such provision of health care was helping anyone in such bodies. The advent of the capacity to transplant organs, in particular, non-paired organs, made it important to determine when one was stealing organs from living persons versus salvaging organs from the alive bodies of dead persons. The fact that we can now sustain the bodies of individuals who will never again be conscious in this world is now forcing us to move beyond whole-brain definitions of death towards higher-brain-center-oriented definitions. We must as a culture now determine, as a matter of law and public policy, whether stopping all treatment including the provision of water and food to a body that will never again sponsor consciousness should count as murder or simply the cessation of support for the body of a dead person.

The recent report of the President's Commission regarding the definition of death does not help us greatly in coming to terms with the implications of the good news-bad news story that supported a higher-brain-centered-oriented concept of death. Indeed, given the central importance of sentience to any notion of persons as objects of respect, the report is somewhat puzzling. In fact, the definition of death forwarded by the President's Commission is remarkably vitalist for a 20th century document. The Commission rejects the notion that what is important about the brain is that it is the sponsor of consciousness ([6], p. 38), and instead underscores the importance of the death of the brain stem as a necessary condition for the death of persons.

"Uniform Determination of Death Act"
An individual who has sustained either (1) irreversible cessation of circulatory and respiratory functions, or (2) irreversible cessation of all functions of the entire brain, including the brain stem, is dead. A determination of death must be made in accordance with accepted medical standards ([6], p. 2).

The Commission offers two interpretations of the significance of its whole-brain-oriented definition of death, each of which, the report suggests, is a mirror image of the other. "The first focuses on the integrated functioning of the body's major organ systems, while recognizing the centrality of the whole-brain, since it is neither revivable nor replaceable. The other identifies the functioning of the whole-brain as the hallmark of life because the brain is the regulator of the body's integration" ([6], p. 32). In accounting for their definition of death, no endorsement, much less emphasis, is given to the centrality of the life of persons as sentient entities. Instead, the emphasis falls on integrated biological functioning. The President's Commission supports the peculiar view that one should be worried about continued integrated biological life even when there is none of the sentience that is a necessary condition for the life of persons.

The cumbersomeness of the proposed uniform statute reflects an interest to eschew simpler definitions that more clearly underscore the uniqueness of the brain, even if those definitions had not emphasized directly the importance of the brain as the sponsor of consciousness. For example, the American Bar Association in 1975 proposed the following definition. "For all legal purposes, a human body with irreversible cessation of total brain function, according to usual and customary standards of medical practice, shall be considered dead" ([6], p. 117). Such a definition without the kind of commentary provided in the President's Commission report is preferable. One might interpret the emphasis on whole-brain as reflecting a concern to avoid false-positive determinations of death that might be associated with a higher-brain-center-oriented definition of death. Moreover, one might even imagine courts redefining the legal understanding of "whole-brain" so as to exclude the brain stem as intermediary structure between the spinal cord and the brain per se. One might speculate that the President's Commission's defense of a vitalist definition of death, underscoring the importance of the brain stem, may have been tied, at least indirectly, to the abortion debate. This debate has also turned on a distinction between the life of persons and mere human biological life ([4], pp. 202–249), [8]. The suggestion that what is important about human life is the capacity of the brain to sustain consciousness would imply that early abortions could not be considered taking a person's life.

The points made by Puccetti are likely to be inescapable if we remain a peaceable secularist society. Economic interests will press us to clarify further what we mean by death. I do not mean to confuse the question of when one ought to sustain the life of persons with the question of when persons no longer exist. Rather, the capacity of our technology to keep biologically alive bodies in which no person lives will force us to address the second question forthrightly on its own merits. In the end, it will be very difficult to defend in general secular terms the view that mere human biological life has an intrinsic moral standing on a par with that of persons. It is likely for such reasons that the President's Commission took back in 1983 some of what it gave in 1981, from the standing of human bodies in which persons will never again be conscious. In its report, *Deciding to Forego Life-Sustaining Treatment*, the President's Commission held that "decisions of patients' families should determine what sort of medical care permanently unconscious patients receive. Other than requiring appropriate decisionmaking procedures for these patients, the law does not and should not require any particular therapies to be applied or continued, with the exception of basic nursing care that is needed to ensure dignified and respectful treatment of the patient" ([7], p. 6). Though the earlier President's Commission Report would not allow one to declare such individuals dead, one may with the permission of the family, at least according to this recommendation, stop all medical treatment. This would appear to include no longer providing antibiotics and, presumably, artificial hydration and nutrition which are medical procedures.[2] This interpretation has recently been endorsed by the American Medical Association [2], though it has been challenged, for example, by the case of *Brophy* v. *New England Sinai*

Hospital [1]. It appears that bodies that have permanently lost consciousness will soon fall into a strange limbo. To shoot them would be murder. One cannot declare them dead and stop treatment. But, according to the Commission, one may stop treatment so that they can then meet the criteria of being dead according to the President's Commission's Uniform Determination of Death Act.

Such bodies can also be the focus of major malpractice suits. Millions of dollars may be awarded for the care and preservation of such bodies, though there is no consciousness, no sentience, no one in the body to benefit from the life continued. In addition, if the awards are not paid out over time, then not only will plaintiff's lawyers reap a large reward, but, if the next-of-kin discontinue all treatment after the award so that the body can die, then the heirs will receive a considerable inheritance. Such outcomes are possible because the legal system has only in part come to terms with what it is to be alive in the world. Until we more clearly face what it is to have died, we will be exposed to such peculiar consequences of false-negative determinations of death. The financial, psychological, and social consequences of such false-negative determination will provide an impetus for change.

We are on a conceptual journey, the beginnings of which Martin Pernick has sketched. Ideas and technology interplay. Their interplay is very likely to lead us to accept higher-brain-oriented definitions of death, such as those for which Puccetti argues. There will surely be, and surely should be, concerns about what tests should be relied on and when. However, as we become clearer about what it means to be we who are here as persons (this is a philosophical and cultural task) and about what it means to be embodied (this is a task for both medicine and philosophy) and about what tests are reliable (this is a task for both medicine and public policy), we will develop ever more carefully focused definitions of death. This will move us beyond whole-brain definitions of death. Already there are proposals to consider anencephalic infants brain absent and surgeons have used such infants as sources of organs [6]. Our ability to do more will force us to be clearer about the significance of the life we sustain. Conceptual distinctions that we could avoid in the past will be pressed on us for recognition.

Center for Ethics, Medicine, and Public Issues
Baylor College of Medicine
Houston, Texas

NOTES

[1] I have explored the issue of the definition of death at some length in [3] and [4].
[2] States such as Oklahoma have attempted to prevent such choices by next-of-kin.

BIBLIOGRAPHY

1. *Brophy* v. *New England Sinai Hospital*, 85E0009–G1 (Mass. Dist. Ct. October 21, 1985).
2. The Council on Ethical and Judicial Affairs: 1986, 'Withholding or Withdrawing Life Prolonging Medical Treatment', American Medical Association, Chicago.
3. Engelhardt, H. T., Jr.: 1975, 'Some Persons are Humans, Some Humans are Persons, and the World is What We Persons Make of It', in S.F. Spicker and H. T. Engelhardt, Jr. (eds.), *Philosophical Medical Ethics: Its Nature and Significance*, D. Reidel Publ. Co., Dordrecht, Holland, pp. 183–194.
4. Engelhardt, H. T., Jr.: 1986, *The Foundations of Bioethics*, Oxford University Press, New York.
5. Holzgreve, W. *et al.*: 1987, 'Kidney Transplantation From Anencephalic Donors,' *The New England Journal of Medicine* **316**, 1069–1070.
6. President's Commission for the Study of Ethical Problems in Medicine and Biomedical and Behavioral Research: 1981, *Defining Death*, U.S. Government Printing Office, Washington, D.C.
7. President's Commission for the Study of Ethical Problems in Medicine and Biomedical and Behavioral Research: 1983, *Deciding to Forego Life-Sustaining Treatment*, U.S. Government Printing Office, Washington, D.C.
8. Puccetti, R.: 1983, 'The Life of a Person', in W. Bondeson *et al.* (eds.), *Abortion and the Status of the Fetus*, D. Reidel Publ. Co., Dordrecht, Holland, pp. 169–182.

PART II

BEYOND WHOLE-BRAIN CRITERIA OF DEATH: LEGAL CONSIDERATIONS

PATRICIA D. WHITE

SHOULD THE LAW DEFINE DEATH? – A GENUINE QUESTION

Until recently there was very little reason for the law to address the determination of death. The determination of death was not itself an issue – it was easy enough for an appropriately trained person, and even for most laymen, to distinguish a dead man from a living one. Is he breathing and is his heart beating? In retrospect, of course, it might be difficult to ascertain the precise moment of death but on the whole not much hinged on such a finding. Now that medical technology has developed sophisticated life support systems, it has produced a class of patients whose status as alive or dead cannot readily be resolved by traditional tests. This fact has made the determination of death an issue for medicine. I would like to explore some of the parameters of that issue and to examine the extent to which it is properly one for the law to address.

There are many legal consequences of death. Many of them involve property rights: the succession of the decedent's property; the entitlement to his or her life insurance and/or pension benefits; the lapse of medical and other insurance coverage. Other consequences affect the legal status of a decedent's spouse and children or other wards. Then, too, there are the inevitable tax consequences. Homicide laws require the victim's death and may depend on how closely the death follows the assault. The nature and extent of tort liability can depend on whether there *is* a decedent. Finally, there are consequences that have a direct bearing on the physical treatment of a person who has been declared dead: the disposal of the body; the availability of various organs for transplant; and the end of the physician's (and others') legal obligation to provide care.

The various legal consequences that flow from death were not established with the current situation in mind. Now that the possibility of forestalling death (in the simple old-fashioned sense) is upon us, it might well be that some of these legal consequences should be looked at again to see if other forms of permanent incompetency should also trigger them. Perhaps, for example, a spouse should be entitled to remarry without having to go through the process of divorcing a spouse in a

Richard M. Zaner (ed.), Death: Beyond Whole-Brain Criteria, pp. 101–109.
© 1988 *by Kluwer Academic Publishers.*

persistent vegetative state. More interestingly, perhaps, as some have suggested, the homicide laws ought to apply (under a different rubric, of course) when the victim is left in a persistent vegetative state. It is important to recognize, however, that a willingness to reexamine the use of death as traditionally conceived as the key to the application of a variety of legal rules does not necessarily commit anyone either to the notion that the law must set forth the conditions of death or to the view that "death" should be redefined.[1]

This straightforward observation about the law leads to a somewhat less straightforward observation about the general issue of defining death. We attach enormous importance to the determination of death. One's own death seems nearly inconceivable in prospect and the finality of another's death is often staggering. The nonlegal consequences of someone's having died cannot be enumerated as easily as the legal ones can, but their significance is even greater. Because death is so significant, the pressure to specify its conditions precisely has grown as medicine has developed the ability to sustain circulation and respiration mechanically. Not surprisingly, concern arose about the status of "brain dead" patients whose single signs of life are ventillator-maintained respiration and a mechanically sustained heartbeat. These are patients who would clearly be dead without their life support systems. With their life support systems they do not, however, satisfy the traditional standards for determining death.

Largely in response to the problems engendered by the ambiguity of the status of such patients,[2] various medical and legal groups and individual scholars have undertaken to examine the adequacy of the traditional standards and to make recommendations concerning the standards that ought to be applied by medicine and the law to determine whether a person is dead or alive (e.g., [5, 6, 4, 14, 8, 10, 13]). In addition, numerous state courts have found themselves presented with questions which have more or less invited them to rule on the adequacy of the traditional standards as determinative of death (e.g., [1, 2, 3]). Although there have been variations in the assessments of those who have addressed the issue of the acceptability of what I have been calling "the traditional standards," the inquiry almost invariably proceeds on the assumption that it has an answer – i.e., that there is a point at which death occurs, that the traditional standards either reflect that point or they do not and that brain-dead patients with artificially maintained circulation and respiration are either dead or alive (e.g., [7, 14]).

However, just as it is possible to contemplate "homicide" laws that apply equally to those who leave their victims clearly dead and those whose victims are left in some more ambiguous state, it is possible to imagine that "death" is not susceptible of precise definition.

The suggestion is perhaps best illustrated by a horizontal line. At one end of the line is death and at the other end is life. Most examples fall clearly at one end of the line or the other. It is almost always obvious whether someone is dead or alive. But there is a class of cases, newly evolved for the most part, which do not fall neatly on one side or the other. Why must we imagine that there is a point on the line which divides the two? Although their logic is somewhat different, it might be useful to think of the task of discovering the point of division between life and death as analogous to what logicians call a "sorites problem." For example, we can all clearly agree that a man with no hair what-soever on his head is bald and that Sampson, at least in earlier life, was not. What about a person with one hair, with two? Is there some specific number of hairs that distinguishes a bald person from one who is not bald? The problem is evident. And, importantly, we are all quite prepared to acknowledge the impossibility of specifying precisely how many hairs one may have and still be bald or how many pieces of straw it takes before a pile of hay becomes a haystack, or how many grains of sand are required to make a beach. We are all quite prepared to regard these as curious features of our conceptual landscape because nothing very important is likely to hinge on the determination of baldness, or status as a haystack, or as a beach.

Death of course is another matter. An enormous amount seems to be at stake in the determination of whether someone is dead or alive. Framing the question in that way, however, is misleading. The instances that force the issue, the situations where there can be any question whether the person is living or dead, are instances where the person himself has comparatively little at stake, for they are instances where the person has permanently lost all cognitive life. But whatever one's view of the gravity of the consequences of a declaration of death to a patient of this sort, the consequences to his family, friends, and medical caretakers *are* likely to be great.[3] Yet the significance of the conse-quences does not entail that there *is* a set of criteria that together are sufficient to distinguish all cases of death from life anymore than significant consequences (e.g., capital punishment for all bald men or an enormous tax on the owners of haystacks) would entail that there is an answer to a sorites problem. It seems to me mistaken, therefore, to *assume* that there is a specifiable point of division between life and death and that all hard cases must fall on one side or the other.

If, as I am inclined to think, there is no one crucial point on the continuum of life and death that separates one from the other, there is a further observation to be made about what is involved in the classification of unclear cases as instances of either life or death. When the question is asked, for example, whether a patient in a persistent vegetative state is alive or dead, what is really being asked is whether that patient is to be treated *as if* he were alive or *as if* he were dead. Conceiving of this question in this way is significantly different from conceiving of it as it has generally been asked by those who have addressed it. For one thing it makes the argument of some philosophers and others that a patient in a persistent vegetative state has ceased permanently to have certain fundamental human characteristics and therefore no longer has status as a person, relevant to the central question in a way in which many critics have thought that it was not (see [11], [10] and contrast [14], [13]). At the same time, though, it makes it clear that that relevance is not necessarily determinative of the question. The argument is relevant because the question asks for reasons to be given for treating a patient as if he were dead, and clearly his permanent incapacity to experience any of the cognitive life that marks human existence is arguably a reason to treat him as if he were dead. The argument, even if valid, cannot necessarily be determinative, however, once one has accepted the hypothesis that there is no critical point dividing death from life and has framed the inquiry in terms of giving reasons for treating a patient as if he were dead or as if he were alive.

Even if we accept, as a theoretical matter, the picture that I have suggested of death as sometimes indeterminate, we must still address the essentially practical issues of how medicine and the law should establish that someone has died. These are tremendously important issues with real consequences and cannot be finessed by the wave of a metaphysical wand. Doctors must make pronouncements of death and there are legal consequences to death. In part, at least, because there are legal consequences to death and in part because there can be legal consequences to doctors for misdiagnosis or other mistakes, it is important that the medical standards that govern physicians not be inconsistent with the standards applied by the law with respect to the determination of death. It is a further step, however, and one which does not seem to me to be clearly warranted, to assert that this need for consistency requires that the law undertake to articulate specific standards for determining death.

Let me try to develop this last point a little more. If, for example, the law were to undertake, by statute or otherwise, to set out that only a patient whose heart and lungs have irreversibly ceased to function is dead, it would have adopted a rather determinate view of what is meant by "death." It would, at the same time, have stopped short of trying to prescribe the methods by which a physician might determine that a patient had died. Thus, such a statute would allow medicine the flexibility to develop new tests to measure the extent of damage to a patient's heart and lungs, but it would require that all determinations of death ultimately be able to be cast in terms of the permanent cessation of heart and lung function. This requirement would surely be uncontroversial if we knew that death was, in principle at least, always determinable and people in general felt comfortable with the point of division articulated by law between life and death. The requirement might also be warranted – although perhaps controversial – if we knew that death was, in principle at least, always determinable and that there was good evidence (whether or not widely understood) for the view that the point of division articulated by law was in fact the point of division. But even if we knew that death was in principle always determinable and if there were serious dispute about where the point of division between life and death is, it seems questionable whether we ought to use the law to inhibit the contribution to that debate that medicine might naturally make if doctors were not forced, at least for prudential reasons, to conceive of the issues within the framework dictated by the law. And if we *doubted* that death was always determinable in principle, a law that articulated a determinable standard could only be justified on grounds of the public policy advantages of using the law to establish and enforce the parameters within which physicians must approach determinations of death. Given that the choice of a determinable standard would, under this hypothesis, be in some sense discretionary, the arguments would need to be made, first, that it was wise to adopt a fully fixed standard and, further, that the institutionalization of that choice in the law was appropriate and desirable.

Another way to achieve consistency between the medical standards that govern the determination of death and the legal ones is to require, for legal purposes, merely that death has been determined by competent medical authority in accordance with the applicable medical standards. This is the relationship that is characteristic of medical and legal standards generally. The law typically is unconcerned with particular diagnoses or treatments *per se*. However, it is relatively rare that significant legal consequences hinge on a particular diagnosis or treatment. In thinking about the appropriate role of the law in articulating standards for the determination of death, one should focus specifically on the question whether the existence and significance of the legal consequences of death justify using the law to set parameters within which medicine must approach determinations of death. If they do not, then, if what I have suggested is right, perhaps the legal standards for determining death *should* be wholly tied to competent medical judgment.[4]

The genesis of this discussion is, of course, the fact that patients can show various degrees of irreversible brain damage. We need to know whether these people are dead or alive and, sometimes, when they died. The mandate of the President's Commission for the Study of Ethical Problems in Medicine and Biomedical and Behavioral Research was to study "the ethical and legal implications of defining death including the advisability of developing a uniform definition of death" ([14], p. 1). The Commission was operating in a context within which (1) there were (and still are) explicit and specific legal standards for determining death and (2) the clear expectation was that a proposal or proposals would issue which undertook to update and make more uniform the various formulations of those standards. It is not surprising, therefore, that the report which accompanies its proposed Uniform Determination of Death Act shows little evidence that it considered seriously the possibility that it should instead have fashioned a systematic retreat of the law from the business of defining death.

An inquiry that proceeds more or less on the assumption that the law has a role in formulating the standards that signify death could proceed to address the specific issue of the role of the neocortex in determining death along one of two general lines. It could, as I suggested earlier, give arguments about the nature of human life and the role that the neocortex plays in any meaningful human existence, and conclude that the law ought (or ought not) to declare that a patient whose neocortex has irreversibly ceased to function is (or is not) dead. If I am right in my suspicion that death is in some sense indeterminate and if such a patient fell within the class of indeterminate cases, this would amount to asking the law to stipulate an answer to what seems a distinctly moral question: i.e., ought we to treat a patient whose neocortex has ceased entirely to function as if he were dead or as if he were alive? Alternatively, such an inquiry could resist, as the President's Commission's Report does, the use of the law to answer that question by distinguishing between the question when someone dies and the question whether someone ought to be allowed to die ([14], p. 4). This distinction which, so far as I can tell, was first emphasized in writing by Capron and Kass in their important 1972 paper (e.g., [8]), is now frequently relied on to justify a legal definition of death that does not provide that the death of the neocortex is tantamount to the death of the person. It may be that patients who have suffered total and irreversible neocortical dysfunction do not fall within the class of unclear cases of life or death, although I am inclined to think that they do. It is important to recognize, however, that if they do, we cannot avoid using the law for moral stipulation by distinguishing between cases of death and cases where passive (or even active) euthanasia might be appropriate. Once the law defines death at all, it effectively mandates that all cases falling outside the boundaries of the definition are not cases of death. Once a doctor must frame an issue in terms of the legality of his allowing a living patient to die, his already heavy moral burdens are increased significantly.

Although my instinct is to be skeptical, I do not yet have fixed views about the extent to which the law ought to be used to set standards for determining death. For the reasons rehearsed here, however, I take very seriously the question whether the law ought to have an *independent* role. I hope in this paper to have convinced others that the question is a genuine and important one.

Georgetown University Law Center,
Washington, D.C.

NOTES

[1] This point is well understood by Capron and Kass [8]. It does not seem to be by Dworkin [9].

[2] Among the most disturbing of these problems are the financial ones. The cost of maintaining a patient on life support systems can be enormous.

[3] For a striking journalistic account of some of these consequences, see [12].

[4] I am *not* suggesting that the law should be uninvolved in specific determinations of death. Instead, I am concerned here to suggest the possibility that it should not establish the standards by which such determinations are to be made. This distinction can be illustrated by contrasting the statements of two judges in two highly publicized court cases – *In re Conroy* [1] and *People* v. *Barber* [2].

In *Conroy* the Superior Court of New Jersey Chancery Division was asked by the guardian nephew of an 84-year-old woman who suffered from severe organic brain syndrome (as well as from a variety of other ailments) for authority to have the nasogastric tube that sustained her removed. Judge Stanton granted this permission and, in the exceptionally thoughtful opinion that accompanied his ruling, wrote:

As often as possible, the patient, the family, and the physicians involved should make these decisions for themselves.

However, fairly frequently judicial involvement is necessary. Sometimes the patient is competent and has not prior to her incompetency given any clear indication of what her desires might be. (This is so in the present case.) Sometimes the family is divided in its views. Sometimes physicians differ among themselves or with members of the family. (This is so in the present case.) When one or more of these factors are present, judicial involvement is indicated.

It should also be noted that the kind of medical ethics committee envisioned by the *Quinlan* case as being available in the typical hospital is not, in fact, in place in many New Jersey hospitals. Such a committee is not available in the typical nursing home. Thus, the kind of solid private institutional support and monitoring of decisions contemplated by the New Jersey Supreme Court in *Quinlan* is frequently not a reality. This means more judicial involvement than would otherwise be the case.

I might also note that I would have some misgivings about a plaintiff such as the present one making basic decisions about termination of treatment without being subject to some kind of judicial scrutiny. Mr. Whittemore is an intelligent and decent man. He is the legal guardian of the patient. He certainly means well for the patient. I believe that in this case he has, in fact, reached the right decision about the nasogastric tube. However, he is only a nephew of the patient, and is, thus, not a particularly close relative. He does not stand in the same relationship to her as would a parent, a spouse, a sibling or a child. Hence, his views are perhaps somewhat less relevant than would be those of a closer relative.

There is a need for some public monitoring of the trend of decisions in this area. Physicians have a technical expertise, a frequent contact, and a professional moral sensitivity which entitle their views to great deference. However, they do not have the public perception and the public responsibility which courts have. Judicial involvement from time to time is, I think, helpful to the integrity and validity of decision making in this area ([1], pp. 11–13).

People v. *Barber* was a murder action brought against two physicians, Drs. Barber and Nedjl, who, with the family's encouragement, removed all of the life support systems (including the I.V. through which he received all of his nourishment) from a patient who had suffered severe brain damage. At one stage in the proceedings (in which the doctors were eventually acquitted), the Superior Court for the State of California reversed the earlier decision of a magistrate to dismiss the charges. In reinstating the charges and requiring the defendants to stand trial, Judge Wenke wrote:

The morality of the defendants' conduct, the purity of their motives, common practice in this type of situation, and the wishes of the decedent's family are all of no weight in the resolution of this motion. The answer lies in the law of this state.

The law of this state does not allow anyone to shorten another's life unless the latter's condition is "irreversible" as hereinabove discussed. The magistrate in effect found that the acts of the defendants shortened the life of the victim and he failed to find that the victim's condition was irreversible ([2], p. 4).

The sort of judicial role described by Judge Stanton is fully consistent with my view. That described by Judge Wenke is not.

BIBLIOGRAPHY

Cases
1. *In re Conroy*, 188 N.Y. Sup. 523 (1983) (decided Feb. 2, 1983).
2. *People* v. *Barber*, No. A 025586 (L.A. Sup. Ct., May 5, 1983) (tentative decision).
3. *In re Welfare of Colyer*, 660 Pac. 2d 738 (Wash. 1983).

Articles and Reports
4. American Bar Association: 1978, 100 A.B.A. Ann. Rprt. 231–32, reprinted in President's Commission for the Study of Ethical Problems in Medicine and Biomedical and Behavioral Research, *Defining Death*, U.S. Government Printing Office, Washington, D.C. 1981, p. 117.
5. Ad Hoc Committee of the Harvard Medical School to Examine the Definition of Brain Death: 1968, 'A Definition of Irreversible Coma', *Journal of the American Medical Association* **205**, 337–340.
6. American Medical Association: 1979, 'An Act to Provide for Determination of Death', reprinted in President's Commission for the Study of Ethical Problems in Medicine and Biomedical and Behavioral Research, *Defining Death*, U.S. Government Printing Office, Washington, D.C., 1981, p. 117.
7. Bernat, J. L., Culver, C. M., and Gert, B.: 1982, 'Defining Death in Theory and Practice', *The Hastings Center Report* **12**, 5–9.
8. Capron, A. M. and Kass, L. R.: 1972, 'A Statutory Definition of the Standards for Determining Human Death: An Appraisal and a Proposal', *University of Pennsylvania Law Review* **121**, 87–118.
9. Dworkin, R. B.: 1973, 'Death in Context', *Indiana Law Journal* **48**, 623–639.
10. Engelhardt, H. T., Jr.: 1975, 'Defining Death: A Philosophical Problem for Medicine and Law', *American Review of Respiratory Disease* **112**, 587–590.
11. Green, M. B. and Wikler, D.: 1980, 'Brain Death and Personal Identity', *Philosophy and Public Affairs* **9**, 105–133.
12. Kleiman, D.: 1985, 'Changing Way of Death: Some Agonizing Choices', *The New York Times* (January 14, 16, and 18).
13. Pallis, C.: 1983, 'Whole-Brain Death Reconsidered – Physiological Facts and Philosophy', *Journal of Medical Ethics* **9**, 32–37.
14. President's Commission for the Study of Ethical Problems in Medicine and Biomedical and Behavioral Research: 1981, *Defining Death*, U.S. Government Printing Office, Washington, D.C.

DAVID RANDOLPH SMITH

LEGAL ISSUES LEADING TO THE NOTION OF
NEOCORTICAL DEATH

(After consuming the bottle labeled "DRINK ME" and shrinking to a height of ten inches, Alice contemplates her fate):
"[S]he waited for a few minutes to see if she was going to shrink any further: she felt a little nervous about this; 'for it might end you know,' said Alice, 'in my going out altogether, like a candle. I wonder what I should be like then?' And she tried to fancy what the flame of a candle is like after it is blown out, for she could not remember ever having seen such a thing."

L. Carroll, *Alice's Adventures in Wonderland* [55].

All human beings know the fact and reality of death, yet like Alice, our existence is in many ways a search to understand death. As Samuel Johnson observed to Boswell, "the whole of life is but keeping away the thoughts of it." Theologians, philosophers, and poets struggle with the meaning of death. Saint Augustine asks what type of death did God intend to enforce His commands. "Was it the death of the soul, or of the body, or of the entire man, or the so-called second death?"[1] Yet for Dylan Thomas, "After the first death, there is no other." The law, too, must face death, particularly the dilemma created by advances in medical science that permit the artificial maintenance of heart, lung, and nourishment functions. This article examines the law's approach to death by inquiring into the legal issues raised by cardiopulmonary, whole-brain, and neocortical definitions of death. The term "cardiopulmonary death" means the irreversible cessation of heart and lung functions.[2] "Whole-brain death" means the irreversible cessation of all functions of the entire brain, including the brain stem. "Neocortical death" means the irreversible loss of consciousness and cognitive functions.[3] Several arguments and a proposal emerge from this inquiry. The law should and does define criteria for death. Unlike Professor White, I believe a contextual approach to death does not offer any significant benefits for medicine or law. Indeed, the vast majority of states either by statute or judicial decision now supplement the cardiopulmonary test for death with a whole-brain death standard. Although widely accepted, the whole-brain definition of death is itself mortal – a

Richard M. Zaner (ed.), Death: Beyond Whole-Brain Criteria, pp. 111–144.
© 1988 *by Kluwer Academic Publishers*.

creation of prevailing medical technologies and a conservative death orthodoxy.

This article carries the legal analysis of death one step further by arguing that neocortical death should likewise be considered as defining the death of the person for all legal purposes. The law now gives *de facto* recognition to the validity of a neocortical definition of death by upholding private decisions to withhold or terminate treatment and nourishment of incompetent, irreversibly non-cognitive patients in a persistent vegetative state. Moreover, recent state-of-the-art medical developments in positron emission tomography (PET) scanning provide further impetus for including neocortical death within the legal definition of death. Although irreversible loss of consciousness and cerebration should establish legal death, the deceased (by a prior written directive) or the family of the deceased should have the option of maintaining biological existence, subject to the financial ability of the estate or family to shoulder the costs of biological maintenance.

I. THE DEVELOPMENT OF THE WHOLE-BRAIN DEFINITION OF DEATH

Death triggers important legal consequences. A determination of death ends marriage and business partnerships, begins the process of disposing of a deceased's property, and may signal the obligation of a life insurance company to pay death benefits or the right to remove the deceased's donated organs for transplantation. Criminal liability for homicide depends on the death of a person. Given the significance of death as a condition precedent to a wide array of legal rights and results, one would think it desirable for law and medicine to formulate a precise conception of when death occurs and what the term "death" means. Yet until relatively recently, the law did not undertake to define death in terms that take into account the new age of artificial life-support systems and transplantation of organs.

For example, suppose that a robber shoots a victim in the head. The victim is rushed to the hospital, placed on a respirator and administered medication to maintain blood pressure. The next morning, a neurologist examines the victim and finds an irreversible cessation of all functions of the entire brain. With the consent of the victim's legal guardian, physicians remove the victim's organs for transplantation purposes. The respirator is then disconnected and the victim's breathing and heartbeat stop. Has the robber committed homicide? Were the medical procedures performed by the physicians a superseding cause of death?

The New York Court of Appeals faced precisely these facts and issues in *People* v. *Eulo* [31]. The defendants in this consolidated homicide appeal argued that they had not caused the death of the victim and that under New York law the court should have charged the jury on the cardiopulmonary definition of death. Rejecting these contentions, Chief Judge Cooke, for a unanimous court, held that the defendants were guilty of murder because whole-brain death could be properly included within the legal definition of death. The court held that death occurs upon a medical finding of cardiopulmonary death or, when artificial means of support preclude a determination that cardiopulmonary functions have ceased, upon a determination that the patient has experienced whole-brain death ([31], p. 295). *Eulo* is a recent example of the law's growing need to resolve the conflict between the conventional view of death as the irreversible cessation of heart and lung function and the medical community's growing reliance and acceptance of brain death criteria [13, 34], ([112], pp. 426–428). A substantial number of cases explicitly or implicitly recognize whole-brain death criteria when application of traditional cardiopulmonary standards would produce unjust results.

Responding to the nagging uncertainty surrounding the legal meaning of death as a result of sophisticated life-support technologies, the law now expressly acknowledges cessation of all brain functions (whole-brain death) as a test for death. To date thirty-three states and the District of Columbia have enacted statutes that incorporate whole-brain death into their definitions of death.[4] Seven other states have judicially adopted a whole-brain death definition for all contexts or for the limited purpose of establishing death under homicide law [2, 6, 8, 11, 31, 35–37, 39].[5] No reported case has rejected the concept of whole-brain death when it has been raised. Of the forty states that have updated legal definitions of death, six states define death by sole or primary reliance on a whole-brain death standard as determined by accepted methods of medical practice.[6] Twenty-six states and the District of Columbia apply an alternative test: either the cardiopulmonary or whole-brain definition.[7] The Uniform Determination of Death Act, recommended to the states by the President's Commission for the Study of Ethical Problems in Medicine and Biomedical and Behavioral Research, adopts this approach.[8] Eight other states retain the cardiopulmonary criteria, but provide that, when artificial means of sustaining respiration and heartbeat preclude application of the cardiopulmonary standard, death may be pronounced if there is a finding of whole-brain death.[9]

In *Eulo* [31], New York adopted whole-brain death for cases involving artificial maintenance of cardiopulmonary functions. Numerous groups, including the American Bar Association, the American Medical Association, the American Academy of Neurology, the American Electroencephalographic Society, the National Conference of Commissioners on Uniform State Laws and the President's Commission, have endorsed the Uniform Determination of Death Act's cardiopulmonary or whole-brain-based criteria ([82], pp. 27–28). In short, law, medicine, and society at large have reached a consensus on the appropriateness of defining the death of the person to include the cessation of all functions of the whole-brain.

One may question, as Professor White ably has done, the desirability of a uniform definition, whether it be cardiopulmonary, whole-brain, or neocortical. Is it wise for the law to articulate a uniform definition of death? Should we not instead employ a contextual approach to the legal problems raised by death? [60] Phrasing the question in this manner, however, minimizes the predicament posed by failing to define death and also understates the law's present general acceptance of whole-brain death. Now that forty states and the District of Columbia have defined death to include whole-brain death, one may question whether this effort should be discarded in favor of a contextual, non-uniform approach to death.

Convincing policy reasons support the effort to articulate a precise standard for death. As Professors Capron and Kass [53, 54] and the President's Commission [97] ably argue, failing to define what constitutes death has numerous drawbacks. Uncertainty and confusion about the line between the living and the dead can result in unnecessary or inappropriate treatment, thereby wasting human, financial and psychic resources. Uncertain conclusions about death also undermine the legitimacy and public support for *post facto* disposition of civil and criminal cases. Will it really suffice to apply a heart-lung standard when the legal issue is inheritance of property, but apply a whole-brain death standard when the question is homicide?

Defining death, and including brain-based criteria within the definition, offers several advantages. One key benefit noted by the President's Commission concerns ease and efficiency in transplanting donated organs. Removing vital organs from a whole-brain-dead donor within the limited time frame in which the donor's heart and lung continue to function [47] materially improves the recipient's chances for a successful transplant. The President's Commission noted, however, that the concern over the determination of death rests less with any wish to facilitate organ transplantation than with the need both to render appropriate care to patients and to replace artificial support with more fitting and respectful behavior when a patient has become a dead body ([94], pp. 23–24). A brain-death standard also defeats the arguments advanced by criminal defendants that a physician's performance of a transplant breaks the chain of causation, thereby relieving the defendant of criminal liability.[10]

In short, a contextual approach to death along the lines suggested by Professors White and Dworkin [60] has not gained acceptance. For example, the recently appointed Federal Task Force on Organ Transplantation has recommended the passage of brain-death laws in the few remaining states that have not yet recognized brain death [43].

II. BEYOND WHOLE-BRAIN DEATH; NEOCORTICAL DEATH

While prevailing preferences embrace a whole-brain definition of death, it would be a mistake to assume that law and medicine will not move forward to recognize a higher-brain-neocortical formulation of what it means to die.[11] The balance of this article addresses the issue at the crux of this volume: why the consensus in favor of whole-brain death must ultimately erode. The analysis concludes with and considers the implications of recognizing neocortical death as legal death.

A person may suffer an irreversible loss of consciousness and cognition, the earmarks of higher brain activity, yet still maintain brain stem functions ([56], p. 991), [67, 73]. A patient in this non-cognitive persistent vegetative state would be dead under a neocortical definition. Such a patient would not be considered dead under a whole-brain death standard because the brain stem, the portion of the lower brain that regulates vegetative functions such as breathing, blood pressure, temperature, and neuroendocrine control, would continue to function. Victims of cardiac or respiratory arrest, asphyxiation, stroke, or head trauma suffer neocortical death, but not whole-brain death, if deprivation of circulatory or respiratory functions occurs for a period of time brief enough to spare the relatively resilient brain stem but long enough to cause permanent damage to the cerebrum [57, 92].[12] Vegetative patients who are neocortically dead can remain biologically alive with intravenous feeding and antibiotics for much longer periods of time than patients who have sustained whole-brain death ([76], p. 197). [13] Although cessation of heart and lung functions typically follows whole-brain death within hours or a few days, patients who suffer neocortical or higher-brain death may maintain cardiopulmonary activities for many years [49], [70]. Karen Ann Quinlan's case is the most familiar example of this phenomenon, as are anencephalic infants.[14]

The clinical picture presented by patients who have experienced neocortical death reflects the destruction of critical elements of the central nervous system, leaving the patient in an irreversible condition in which there is no awareness, thought, or feelings ([74], p. 8; [115], p. 256). Deprived of higher brain functions, the patient does not purposefully react to external stimuli. A patient may independently maintain heartbeat and breath and may yawn, sigh, or react to light. These responses, however, are merely physiologic reflexes. As Younger and Bartlett conclude, "Despite the continued ability to spontaneously integrate vegetative functions, a patient who has irreversibly lost the capacity for consciousness and cognition is dead. What remains is only a mindless organism" ([115], p. 256).[15]

Numerous observers of medical ethics echo the view that patients who experience neocortical death and fall into a persistent vegetative state should be treated as brain dead [64–67, 60, 91, 95, 97, 106–108, 115]. The essential argument these commentators present for a neocortical death formulation hinges on the centrality of consciousness and cognition as the quintessential attributes of human life. Jacob Bronowski's elegant study, *The Ascent of Man*, expresses the concept from the standpoint of a scientist and humanist:

Man is a singular creature. He has a set of gifts which make him unique among the animals; so that unlike them, he is not a figure in the landscape – he is a shaper of the landscape . . . Among the multitude of animals which scamper, fly, burrow and swim around us, man is the only one who is not locked into his environment. His imagination, his reason, his emotional subtlety and toughness, make it possible for him not to accept the environment but to change it . . . Man is distinguished from other animals by his imaginative gifts. He makes plans, inventions, new discoveries, by putting different talents together; and his discoveries become more subtle and penetrating . . . [W]hat makes man what he is? . . . How did the hominids come to be . . . dexterous, observant, thoughtful, passionate, able to manipulate in the mind the symbols of language and mathematics, the visions of art and geometry and poetry and science? ([52], pp. 14–30).

Pascal's metaphor of man as a *roseau pensant*, a thinking reed, is perhaps the most vivid articulation of the cerebral quality of human life: "Man is a reed, a bit of straw, the feeblest thing in nature. But he thinks. He is a thinking reed . . . Man's dignity, our dignity, lives in our thoughts. Thereby we rise. Only thereby . . . A thinking reed. Not in space am I to seek my dignity. But in my thinking" ([93], pp. 74, 117; also [58], p. 28).[16] If neocortical or higher brain functions – the capacity to think, feel, communicate, or experience our environment – are the key to human life, then neocortical function should be the key to human death. As Robert Veatch reasons, if death is characterized by the irreversible loss of those attributes of an organism essentially significant to it, and if in humans these attributes are the capacity for consciousness and higher cortical functions rather than for autonomic bodily integration, then patients who have irreversibly lost these distinguishing features of human life should be treated as dead ([108], pp. 312–313). As Senator Jacob Javits, himself a victim of a terminal illness, stated in testimony before a House Select Committee on Aging, "Because medical technology can now sustain life even when the ability to think is gone, society must change its laws" [42].

Reserving judgment for the moment on the philosophical question of whether irreversible loss of higher brain functions constitutes human death, one must still recognize that the expanded time frame of biological existence following neocortical death raises a host of real and immediate problems for law and medicine. Perhaps the most troubling question raised in cases involving patients who have lost higher brain functions is whether withdrawing artificial life-support (respirator and drug therapy regimen) and feeding (intravenous or nasogastric nourishment) is justified. A growing number of cases and statutes now permit withholding or complete withdrawal of life-sustaining treatment and, more recently, nourishment in cases involving patients who have irreversibly lost all cognitive functions. The import of these decisions is profound. By dispensing biological death to irreversibly non-cognitive patients these authorities force us to rethink the appropriateness of a new legal definition of death that includes irreversible loss of consciousness and cognitive functions.

The *Quinlan* case [20] and its progeny[17] recognize that an incompetent patient has a right of privacy, which includes the right to terminate or refuse life-supporting care. Futher, these rights may be exercised by substitute decisionmakers (third parties). Although originating in cases involving irreversibly non-cognitive patients, the right to privacy/right to die rationale has been extended to cognitive incompetent patients who are terminally ill, old, and mentally impaired.[18]

Giving legal recognition to neocortical death, however, could advance the analysis of the sensitive issue of forgoing or withdrawing nourishment and artificial life-support systems in cases involving incompetent terminally ill patients (infants and adults) who nevertheless retain cognitive functions, by forcing physicians and families to focus on the distinction between patients who are conscious and alive, and patients who are irreversibly non-cognitive and, therefore, dead.[19]

The case for redefining death in neocortical terms becomes compelling when one examines the logic and results in the right to privacy and right to die cases. In *Barber* v. *Superior Court* [1], a California Court of Appeals issued a writ of prohibition to bar the prosecution of two physicians on murder charges for discontinuing life-support equipment and intravenous feeding of a irreversibly non-cognitive adult patient.[20] The court held that the physicians' omission to continue to treat the patient at the written request of the patient's wife, though done intentionally and with the knowledge that the patient would die, was not unlawful ([1], pp. 1015–1022). The importance of *Barber* lies in its refusal to apply homicide laws to a case involving the intentionally caused death of a patient who, although neocortically dead, was a living person under whole-brain death law.

In *In re Conroy* [11] the New Jersey Supreme Court ruled, on privacy and self-determination grounds, that an incompetent yet cognitive, institutionalized, elderly patient (83 years old), with severe and permanent mental and physical impairments and a life expectancy of approximately one year, could be disconnected from all life-sustaining treatment and nourishment (nasogastric feeding tube) if such a decision would have been desired by the patient. The decision is significant for two reasons. First, the court explicitly approved cessation of nourishment, equating nasal tube feeding to artificial breathing induced by a respirator ([11], pp. 1233–1237). The opinion rejected any attempt to distinguish between actively hastening death by terminating treatment and passively allowing a person to die of disease. Second, the court authorized termination of all life-support measures in a case involving a terminally ill patient who retained consciousness and limited cognitive functions ([11], pp. 1216–1217) and was, therefore, not brain dead under whole-brain or neocortical standards. In so doing, the court reversed an appeals court ruling [10] that had denied termination on the ground that the right to terminate life-sustaining treatment based on a guardian's judgment was limited to incurable and terminally ill patients who were brain dead (under whole-brain criteria) or irreversibly non-cognitive (neocortical death), and who would gain no medical benefit from continued treatment ([10], p. 310).

Finding a common-law right of privacy and self-determination to decline or terminate medical procedures for both competent and incompetent patients, the New Jersey Supreme Court ruled that incompetent patients should have the right to discontinue life-support and feeding regimens ([11], pp. 1227–1229). Guardians must seek to determine whether the incompetent patient would have desired termination according to three standards: subjective (when it is clear that the patient would have refused treatment under the circumstances involved, e.g., living will or oral statements); limited objective (some trustworthy evidence of what the patient would have desired plus satisfaction on the part of the decision-maker that the burdens of the patient's life with the treatment outweigh the benefits); and pure objective (net burdens of life clearly outweigh benefits and pain makes further treatment inhumane) ([11], pp. 1229–1233). The guardian's decision would be subject to state administrative review ([11], pp. 1241–42). The court held that the evidence at trial was inadequate to satisfy any of the tests and a new trial would have been necessary had the patient, Claire Conroy, lived. Claire Conroy died while the case was on appeal, but the New Jersey courts decided to resolve the issues presented by the case because of their significant public importance ([11], p. 1219).

Cases from the state of Washington further illustrate the growing trend toward acceptance of private decisions to forgo treatment in cases involving patients who have irreversibly lost higher brain functions and are, therefore, neocortically dead. In 1980, Washington accepted whole-brain death as the test for when a person is legally dead in *In re Bowman* [8], [99]. In *Bowman*, the court stated that it was not deciding "the much more difficult question of whether life-support systems may be terminated if a person is in a chronic persistent vegetative state" ([8], p. 735). In 1983, however, the Washington Supreme Court faced this issue in *In re Colyer* and answered in the affirmative ([9], p. 750).[21] The same court reached a similar result in *In re Hamlin* [14], a case that involved an irreversibly non-cognitive patient in a persistent vegetative state who had no family and, unlike the patient in *Colyer*, had been incompetent his entire life.[22]

In a Georgia Supreme Court case, *In re L. H. R.*, the court considered "under what circumstances may life-support systems be removed from a terminally ill patient existing in a chronic vegetative state with no hope of development of cognitive function" ([18], p. 717). *L. H. R.* involved the birth of an infant whom a neurologist described as being in a chronic vegetative state with no hope of development of cognitive function ([18], p. 718). Eighty-five to ninety percent of the infant's brain tissue had been destroyed and her condition was described as irreversible. The court reasoned that the right to refuse treatment could be exercised by the parents or the legal guardian of the infant after a diagnosis that the infant was terminally ill with no hope of recovery or upon a finding that the infant existed in a chronic vegetative state with no reasonable possibility of attaining cognitive function. The court further stated that, in cases involving irreversible loss of cognitive functions, there is no legal difference between the situation of an infant and an incompetent adult who has made no living will. Accordingly, the court extended its holding to cases involving incompetent adult patients who are terminally ill in a chronic vegetative state with no reasonable possibility of regaining cognitive functions ([18], pp. 722–723). Cases in Connecticut, Delaware, Florida, Louisiana, Massachusetts, Minnesota, New York, and Ohio have also sanctioned efforts to withdraw artificial life support machinery from irreversibly non-cognitive infants and adults.[23] Three states have also passed natural death statutes that specifically provide for withdrawal of life-sustaining treatment for patients who have not executed a natural death directive and are diagnosed as unconscious with no reasonable possibility of returning to a cognitive sapient state.[24]

The development of a distinct right-to-die jurisprudence in cases involving cessation of life-support systems for patients who have irreversibly lost consciousness, coupled with the New Jersey Supreme Court's elimination of any distinction between nourishment and other forms of artificial life-support measures in *In re Conroy* [11], have prompted efforts to terminate ill patients who linger in a persistent vegetative state with no hope of regaining consciousness. For example, on February 6, 1985, the wife of a 47-year-old fire-fighter filed a declaratory judgment petition in probate court in Dedham, Massachusetts, to obtain a court order to compel the New England Sinai Hospital in Stoughton, Massa-chusetts, to cease administering food and water to her husband, Paul Brophy [36]. Mr. Brophy suffered a cerebral aneurysm in 1983 and lapsed into irreversible unconsciousness. Neurological evaluation demonstrated he was able to breathe and independently maintain his heartbeat but was totally unaware and incapable of thought or consciousness. The trial of the case lasted eight days and attracted an array of nationally prominent medical experts and ethicists. Expert physicians testified that removing the stomach tube that provided nourishment would cause death in one to two weeks.[25] The filing of the suit attracted national attention because the case was the first time that a court had been asked to order the removal of a feeding tube from a patient who was permanently vegetative and not otherwise terminally ill.

On October 21, 1985, Probate Justice David Kopelman rendered judgment and permanently enjoined The New England Sinai Hospital, its physicians and staff from removing or clamping Mr. Brophy's feeding tube for the purpose of denying hydration and nutrition [30]. The court also permanently enjoined Mrs. Brophy from authorizing any other medical facility from discontinuing nutrition and hydration in the event Mr. Brophy were to be transferred from the hospital to a nursing home. In barring Mrs. Brophy's attempt to obtain judicial approval to discontinue feeding, the court relied on a number of evidentiary and legal factors.

At the outset, the court had some factual misgivings as to whether Brophy was completely unconscious ([30], p. 11). Nevertheless, the court concluded that the patient was in a persistent vegetative state, and that it was highly unlikely that Brophy would ever regain cognitive abilities ([30], p. 12). In focusing on the artificial feeding issue, the court found that the gastrostomy tube method of furnishing food and water posed no serious risks of pain or other discomfort, whereas the possibility of experiencing pain as a result of the withdrawal of food and water could not be ruled out. The court's inability, or unwillingness, to pronounce Mr. Brophy completely non-cognitive arguably provided a sufficient basis to deny the requested relief. If the patient were in a "locked-in" syndrome or could actually think or feel at some meaningful level, the argument for terminating feeding becomes much more troublesome.

The court, however, found a different basis for its decision to enjoin removal of the feeding tube. Even though Brophy previously had conveyed to his family that he did not want to be kept alive artificially should his condition become hopeless ([30], pp. 24–25), these wishes, whether expressed by Brophy himself or on his behalf by his guardian, were outweighed in the court's view by "the most significant state interest of all . . . the preservation of human life" ([30], pp. 39–40). The court held that the fundamental right to refuse intrusive medical treatment must be subordinated to the state's interest in preserving human life when the patient is not terminally ill, has not reached the end of his normal span of years, and the treatment is not highly invasive or painful ([30], pp. 40–42). By contrast, the court stated that were Brophy terminally ill or dying, "it might be permissible to remove the feeding tube" ([30], p. 43). The court also rejected any effort to justify termination of feeding on quality of life grounds. "Otherwise," the judge observed, "the court is pronouncing judgment that Brophy's life is not worth preserving. The quality of life is an incorrect focus because there are no measurable criteria for making such a judgment" ([30], p. 43).

The probate court did not, however, have the final word on Mrs. Brophy's request. On February 14, 1986, the Supreme Judicial Court of Massachusetts *sua sponte* transferred the *Brophy* appeal for hearing by that court without waiting for a decision by the intermediate appellate court.

Justice Kopelman's decision raises several important questions. First, does it make a difference that the request is to discontinue feeding via a tube into the patient's stomach as opposed to a request to discontinue supplying air or blood flow via other machinery? *In re Conroy* [11] and *Barber* v. *Superior Court* [1] (by implication) said no.[26] *Brophy* says yes. The result in *Brophy* is somewhat surprising in that a prior Massachusetts Appeals Court case permitted withholding abdominal surgery from an elderly mentally ill patient who needed the surgery to have a feeding tube inserted in order to be nourished [16]. On the merits of the question it is difficult to conceive how feeding through a stomach tube is qualitatively less artificial or less invasive than other forms of life-support regimens which courts have allowed to be withdrawn from irreversibly unconscious patients.

Second, does it make a difference in dealing with a permanently brain-damaged patient in a persistent vegetative state that the patient suffers from some terminal illness *other* than brain damage? A number of cases outside of Massachusetts have treated irreversible unconsciousness as a sufficient condition in itself to justify terminating treatment [1],

[9] or have characterized irreversible unconsciousness as a terminal illness [18]. *Brophy*, however, made no reference to these cases and endeavored to distinguish prior Massachusetts decisions that had permitted termination of treatment by noting that the prior Massachusetts authorities involved terminally ill (as opposed to vegetative) patients or procedures that, unlike feeding, were extremely invasive.

Third, is a patient in a persistent vegetative state a "human" life so as to call into play the state's paramount interest in preserving human life? This is the key question. Indeed, if the answer is negative the first two questions become irrelevant. *Brophy* did not confront this question directly. Instead, the court simply advanced the state's interest in preserving human life as the principal factor affecting the decision to bar removal of the feeding tube.

In short, although *Brophy* places a misguided emphasis on the supposed differences between artificial life-support systems and artificial feeding, and terminal illness as opposed to terminal unconsciousness, the court nearly hit the mark in focusing on human life. The correct question, though, is not whether the state's interest in preserving human life should prevail over the wishes or substituted judgment of a patient in a persistent vegetative state; rather, is a patient who is doomed to unconsciousness and is no longer aware of the environment a human life?

These cases and statutes involving private decisions to terminate life-support systems of patients in a persistent vegetative state, in tandem with the recent efforts to cease artificial feeding of irreversibly unconscious patients, compel the conclusion that neocortical death is equivalent to death of a person. By holding that third parties, families, or legal guardians may decide to terminate treatment or feeding of an incompetent patient who has irreversibly lost all cognitive functions, the law now gives *de facto* recognition to neocortical death. Either irreversibly unconscious and non-cognitive patients are dead or they are alive. If alive, these patients are being put to death by relatives, guardians, and courts under the logic of substituted judgment or euthanasia.

The final chapter in Karen Ann Quinlan's case vividly illustrates the point. Karen Quinlan was pronounced dead on June 11, 1985. She did not die of any brain disorder, but from pneumonia which had been diagnosed for months. According to the physician responsible for her treatment, the family had asked that no extraordinary measures be taken to keep her alive, including the administering of antibiotics and blood pressure drugs [86]. The Quinlan family had been alerted five days before June 11th that death was imminent but renewed their request that no revival efforts be undertaken [86]. If Karen Ann Quinlan was a living human being prior to June 11, 1985, the family exercised, in a fashion, passive euthanasia by allowing her "to die" when and as she did.

The point here is not one of fault but how best to cope with the angst of decisions concerning permanently vegetative patients. Which analysis for treating the irreversibly unconscious as dead makes more sense: withholding or stopping feeding or life-support therapy because the patients are already dead, or terminating treatment or life-sustaining nourishment of living persons because that is what substitute decision-makers suspect the patients would have wanted?[27] Ordinarily, intentionally ending the life of another, whether by act or omission, is homicide. One can finesse this argument and rationalize the substitute judgment approach by arguing, as courts following *Quinlan* have done, that dispensing death to an incompetent patient is no more than affirming the incompetent patient's constitutional or common-law right to refuse medical treatment and die. This viewpoint may be the only acceptable solution to the dilemma of withdrawal or refusal of treatment or nourishment from living persons who retain consciousness and cognition but suffer from terminal illness.

If neocortical death is the death of a human being, however, the substitute judgment test is an unnecessary mind trip, a profound leap into the dark world of the permanently insentient. Worse, it creates procedural and legal presumptions against withholding or terminating treatment or nourishment which unreasonably burden families, physicians, and courts with the agonizing decision of whether to "play God" and "let the patient die" even though, rightly viewed, human death has already occurred. Finally, the desire to obtain the legal results of death (insurance benefits, inherited property, favorable date of death tax valuations, or remarriage) may motivate relatives or guardians to terminate a patient's biological existence or deliberate precisely as to when death should be doled out.

A more just and sensible position is to consider irreversibly unconscious, non-cognitive patients legally dead, but to recognize and account for the possibility of continuing biological existence. The law should invoke criminal liability and set in motion the other legal consequences that flow from the fact of death when a person suffers neocortical death.

In the earlier example of a robber who shoots a man in the head, suppose that instead of sustaining whole-brain death the victim retains brain stem function but falls into a chronic unconscious vegetative state with no hope of regaining cognitive function. Should the robber escape a murder prosecution?[28] If the victim's spouse is named as a beneficiary under a life insurance policy, should the life insurance company be able to deny death benefits? Should the spouse be barred from remarrying? Should relatives watch the stock market for a timely drop in stock prices to decide when to terminate life-support systems in order to obtain the most economically favorable date of death tax valuation? Little in logic or social policy supports answering any of these questions affirmatively. A death certificate establishing death for all legal purposes should be issued on the date of a medical finding of neocortical death.

Under a neocortical death definition, Karen Ann Quinlan would have been declared legally dead in 1975 instead of 1985. Thousands of other patients who now biologically subsist in a chronic vegetative state with no hope of regaining cognitive function would also be considered legally dead. Would this mean that all persons declared legally dead by neocortical standards would have to be buried or cremated?[29] Would it be permissible for families or guardians to maintain biological existence if they so desired or if the patient had so specified by a pre-existing death directive, or living will?

Treating the neocortically dead as legally dead does not and should not necessarily require burial or cremation. A patient by prior directive, or the patient's family or guardian by a post-neocortical death decision, should be able to provide for and effect biological maintenance notwithstanding a legal certification of neocortical death. Once the law treats neocortical death as legal death, there is no public health need or other strong policy reason to require disposal of the body. In effect, after a legal and medical determination of neocortical death, the law need not act. Burial, cremation, and biological maintenance are all within the realm of appropriate choices. Thus, in the unlikely event that a person wished to be fed and maintained by artificial means in the event of neocortical death, or if the patient's family or relatives wished to maintain the patient's biological existence, these desires could be effected. For persons who desired yet could not afford artificial maintenance, the medical profession and legislators would have to decide whether to subsidize biological maintenance of a neocortically dead body. Although there are certainly grounds for debate on this point, continuing the existence of a legally dead body should be subject to the financial ability of the estate or relatives to pay for the costs of treatment. Society should not incur expenses for medical maintenance because society's interest is minimal, given that the body is legally dead.[30] In this context the tradeoff or cost/benefit analysis is not between dollars and human life but between dollars and biological maintenance.

Absent a directive of the deceased or next-of-kin for biological maintenance, however, burial or cremation would occur. How would this be done? As Karen Quinlan's case demonstrated, stopping life-support systems may not produce cardiopulmonary death if the patient is capable of spontaneous ventilation. To effectuate biological death, two procedures seem logical: termination of nourishment and fluids or active termination by chemical injection. Ending nourishment and hydration produces biological death, but by a slow process of starvation and dehydration. Although a neocortically dead patient does not experience the pain of starvation and thirst, relatives and friends certainly can suffer in witnessing their loved one wither away over a period of weeks. Thus, if a neocortically dead patient biologically subsists without the aid of artificial life-support machines, active termination by injection may be a more humane procedure to bring about biological death than withdrawing fluids and nourishment. As in the case of whole-brain death, however, neocortically dead patients should be buried or cremated only on the cessation of cardiopulmonary functions.

A practical problem may arise. Suppose physicians, nursing home employees, or others charged with caring for neocortically dead patients refuse to withdraw feeding tubes or inject chemicals to bring about biological termination out of fear of criminal or civil liability or because of personal ethical or religious beliefs? Several points may be suggested. First, once neocortical death becomes legally recognized, fears of criminal and civil liability for performing appropriate termination procedures pursuant to the wishes of the patient or family become unfounded. A neocortical death statute should define death, outline termination procedures and incorporate an immunity principle in clear terms. A model statute might read as follows:

NEOCORTICAL DEATH

Sec. 1. As used in this statute, the term "neocortical death" means the irreversible loss of an individual's consciousness and cognitive functions. An individual who has sustained neocortical death is legally dead. A determination of neocortical death under this section must be made in accordance with reasonable medical standards and procedures.

Sec. 2. After a medical determination of neocortical death the individual may be biologically maintained if the individual has executed a written instrument (in accordance with the jurisdiction's living will statutes or procedures) expressing the desire to be maintained on artificial life-support systems. If the individual has made no such prior written declaration, the [family, next of kin, or guardian] may provide for biological maintenance if maintenance is desired.

Sec. 3. If neither the individual (by a prior written directive) nor the [family, next of kin, or guardian] elect to maintain the neocortically dead patient's body, all artificial life-support systems may be withheld and terminated and the provision of nourishment and fluids may be withheld or ceased. As an alternative to the withholding or cessation of nourishment and fluids as a means of terminating biological existence, the [family, next of kin, or guardian] may elect to request injection of a chemical in a quantity sufficient to cause biological death. The chemical must be administered in accordance with reasonable medical procedures.

Sec. 4. No person, firm, or organization shall be subject to criminal responsibility or to civil liability for terminating the biological existence of a neocortically dead individual by any of the methods or procedures authorized in Section 3 [withholding or terminating artificial life-support systems, cessation of nourishment and hydration, or lethal chemical injection].

With legal recognition of neocortical death and an express criminal and civil immunity for terminating the biological existence of neocortically dead patients, one would expect the evolution of a medical and ethical consensus which considered termination of neocortically dead patients by cessation of nourishment and fluids or by chemical injections an acceptable and ethical medical practice. A consensus along these lines is already developing. For example, in July, 1985, the Massachusetts Medical Society's governing council endorsed a resolution that recognizes the appropriateness of discontinuing nourishment in the case of vegetative individuals:

The MMS recognizes the autonomy rights of terminally ill and/or vegetative individuals who have previously expressed their wishes to refuse treatment, including the use of intravenous fluids and gastrointestinal feeding by tube and that the implementation of these wishes by a physician does not itself constitute unethical medical behavior provided that appropriate medical and family consultation is obtained [88].

In short, considering that a respectable body of medical opinion already views termination of feeding in cases involving vegetative patients as an ethically and medically acceptable practice, it is reasonable to assume that with clear legal validation of termination procedures the medical community will generally respect the wishes of the individual patient or family to cease maintaining a legally dead body.

III. DIAGNOSING NEOCORTICAL DEATH

This discussion assumes that neocortical death would be diagnosed in accordance with accepted medical procedures. Arguably, the right-to-die cases involving irreversibly unconscious patients already establish a medical and legal precedent because expert physicians repeatedly have testified with confidence that the patients involved in the particular cases were irreversibly unconscious and insentient.[31] In a great number of cases, particularly when unconsciousness in a vegetative state has lasted for more than one month, physicians routinely diagnose irreversibility based on clinical evaluations and electroencephalogram (EEG) tests. Yet critics of a neocortical death definition repeatedly refer to the difficulty of formulating reliable diagnostic criteria for neocortical death as a major pragmatic stumbling block [47, 4, 94, 113].

While as early as 1971 physicians in Great Britain contended that in most cases a flat electroenecephalogram and biopsy specimen could confirm neocortical death ([51], p. 565), these procedures, unlike well-accepted whole-brain death criteria, are an imprecise and invasive means for diagnosing irreversible unconsciousness. In 1983, for example, Youngner and Bartlett, themselves proponents of a neocortical definition, concluded: "At present, clinicians are unable to apply a definition that identified death as the absence of consciousness and cognition. Medical science has not yet developed tests that accurately establish the irreversible loss of these functions" ([115], p. 258).

Several important observations enter at this point. First, that neocortical death cannot be diagnosed with certainty in all cases does not mean that in a significant number of cases physicians cannot diagnose irreversibility with certitude. At present neurologists are doing so and are testifying to the fact in court cases involving efforts to terminate life-support machinery and artificial feeding. Second, the limits of present medical technologies are no reason to avoid addressing the appropriateness of a neocortical death standard, because we may assume that science will progress. Third, and most important, science is on the verge of doing so. There is very strong evidence that the recent advent of positron emission tomography (PET) scanning now offers the scientific capability of accurately diagnosing metabolic brain function and neocortical death by radioactive scanning.

The rapid advancement of PET scanning technology is a relatively recent occurrence. On January 3, 1985, an editorial by Dr. Henry Wagner, Jr., in the *New England Journal of Medicine* hailed the dramatic development of PET scanning.[32] By using specific tracers labeled with positron emitting isotopes and a video screen, it is now possible to measure the brain's intricate chemistry *in vivo* by viewing blood flow or uptake of glucose and oxygen in selected subregions of the brain and thereby assess higher brain functions. Yellow and green depict normal metabolic activity; blue or purple indicates low activity or no uptake.[33]

As greater numbers of neocortically dead patients undergo PET scans, medical scientists will be able to formulate levels of bio-energetic chemical utilization below which persons do not ever regain consciousness. In short, with the breakthrough in PET scanning, it is no longer tenable to argue that neocortical death is incapable of being reliably diagnosed.

IV. MEDICO-LEGAL IMPLICATIONS OF NEOCORTICAL DEATH

Legal recognition of neocortical death raises numerous questions and issues, perhaps the most intricate of which concerns transplants. A neocortical death standard could significantly increase availability and access to transplants because patients (including anencephalics) declared dead under a neocortical definition could be biologically maintained for years as opposed to a few hours or days, as in the case of whole-brain death.[34] Under the present Uniform Anatomical Gift Act, this raises the possibility that neocortically dead bodies or parts could be donated and maintained for long term research, as organ banks, or for other purposes such as drug testing or manufacturing biochemical compounds [45]. At the outset one should note that the transplant issue in cases involving neocortical death under the legal death/biological death differentiation advanced in this article arises only when the patient or patient's family elects not to preserve biological existence. If either the patient or family directs that biological existence be maintained after neocortical death, transplantation or other use of the body by science prior to cardiopulmonary death by virtue of a donation card or document signed by the deceased is inconsistent with the desire to preserve biological existence. In this event, what effect should be given to a donor card or document signed by the deceased? Or if the family decides not to maintain biological existence, does this preclude the family from donating the neocortically dead body so that scientists or physicians may biologically maintain the body for transplant or research purposes?

The prospect of keeping a donated body biologically alive for years for scientific purposes can be dealt with in several ways. One tack is to maintain the status quo. Under the existing procedures set out in the Uniform Anatomical Gift Act, the decedent or relatives could donate the neocortically dead but biologically alive body or parts for "medical or dental education, research, advancement of medical or dental science, therapy, or transplantation."[35] Long-term research or transplant operations would depend on medical ethics. This result can be supported by arguing that there is no qualitative difference between the types of procedures, transplants, or research that are currently allowed on bodies that have been declared dead under whole-brain standards but which still maintain cardiopulmonary functions, and the types of research, transplants, and experiments that would be undertaken on bodies that have been declared dead under neocortical criteria but continue to exhibit cardiopulmonary activity. Only the time window for action has been extended by a neocortical standard.

Yet the possibility of removing the eyes or a kidney from a body that breathes and has a heartbeat – something that could be done under present whole-brain standards – and then maintaining the body for years for other transplants or research – something that is possible only under a neocortical approach – may call for reappraising the Uniform Anatomical Gift Act's transplant procedures. The status quo approach also would not necessarily alert families or persons who sign donor cards that a donation might result in the corpse being used for human experimentation or long-term scientific research. Therefore, in cases involving neocortical death in which the body or parts thereof were intended for long-term research, transplants, or experiments that would not have been possible under whole-brain circumstances, one might forbid third-party donations and require a written donation document signed by the decedent that donated the entire body or relevant organs. Or, more narrowly, one could require a donation consent form that described the chronic persistent vegetative state-neocortical death circumstance. An intermediate position would permit third-party donations for long-term maintenance for transplant or other scientific purposes if the family or guardian were fully informed of the nature of the intended use of the deceased's body or organs.

These alternatives assume the ethical propriety, in at least some circumstances, of keeping a neocortically dead body alive for some medical or scientific purpose.[36] Yet, is society ready to seek and accept donations of neocortically dead bodies in order to carve out cartilage now and then or to inject such bodies with the virus responsible for AIDS in hopes that experimentation will yield a cure? If not, then we should treat neocortically dead bodies as we now treat whole-brain donations: biological existence should be maintained only for limited transplant purposes with the view toward terminating cardiopulmonary functions following the transplant procedure.

Giving legal effect to neocortical death does not present any special legal problems that cannot be resolved with just results. In the area of tort liability for causing a person to lapse into an irreversibly non-cognitive condition, the law currently subjects the tortfeasor to liability for personal injuries via an action for damages. If the victim were to be declared neocortically dead, no real change in the tort law need take place. The tortfeasor would be subject to liability for wrongful death under survival and wrongful death statutes. It is conceivable that a court might have to rule on the measure or extent of damages in cases in which the patient or family had exercised the decision to maintain the neocortically dead body's biological existence. Under existing law, the tortfeasor would be liable for such medical expenses. With legal recognition of neocortical death, the defendant might contend that the deceased's estate or family should bear these costs, on grounds that they are avoidable damages because the law would permit termination of treatment in the case of neocortical death. This mitigation of damages theory should be rejected with no change in the tort law. If a family or patient desired maintenance, the tortfeasor should bear this expense, consistent with both the result under the status quo and the principle that the tortfeasor takes the victim as he finds him.

In the criminal law context, another potential scenario involves the possibility of a person who intentionally and wrongfully "pulls the plug" on a neocortically dead patient who is being biologically sustained pursuant to the wishes of the family or the prior directive of the decedent. Because the body is dead in the eyes of the law, the person who stopped the machines would not be guilty of manslaughter.[37] The actor should be civilly liable, however, for the intentional infliction of emotional distress to surviving family members. If criminal sanctions seem more appropriate, legislators are of course free to fashion criminal penalties for such conduct.[38]

A neocortical definition of legal death would also affect insurance law. At present irreversibly unconscious non-cognitive patients continue to receive health insurance and disability benefits, but do not receive death benefits under life or accidential death insurance policies [3], [7], [28]. Under a new legal definition of death that incorporated neocortical death, just the opposite results would occur because the insured would be legally dead. Life and accidental death benefits would be paid to beneficiaries, but insurance companies would not be obliged to pay health or disability benefits under current policies. If the law recognized neocortical death but reserved the right to keep the body biologically alive, it is likely that insurance companies would rewrite or offer new health insurance policies that provide coverage for the costs of biological maintenance following a certification of neocortical death. As a matter of private contract law, insurance companies and persons desiring to insure against the costs of biological maintenance should be free to seek and obtain coverage.

Another case that may arise concerns the potential for a conflict between the desires of the decedent and the decedent's heirs or devisees when the decedent has executed a written directive to be maintained biologically. Suppose that, while competent, the decedent wrote a "living will" that expressed the desire to be kept on life-support systems in the event of irreversible loss of cognitive functions. The cost of such maintenance would have to be borne by the decedent's estate or the decedent's family. If the decedent's heirs or beneficiaries did not desire maintenance, however, they might contend that the estate's assets were being "wasted" in maintaining a dead body. In this situation, the decedent's directive should control. While under present law the directive of deceased persons concerning disposition of their bodies is treated only with "benevolent discretion" by courts [61], a directive to maintain biological life is materially different from a request to be buried with "all my diamonds, stock and sterling silver" [29]. This result is also consistent with the outcome under present law – a patient's assets would be used to pay for maintenance costs if that is what the patient had directed or would have desired.

In summary, while one can think of numerous legal nuances that will or may arise if the law sanctions neocortical death, the perceived problems do not appear to be serious or incapable of fair resolution.

V. CONCLUSION

The current legal treatment of brain death is both anomalous and unsound. The current law is grossly inconsistent: it upholds surrogate decisions to terminate life-support systems and nourishment when incompetent patients irreversibly lose all consciousness and cognitive functions, yet fails to recognize neocortical death. If society accepts sentencing death to the irreversibly non-cognitive – and there is a growing medical, legal, and public consensus to do so – then it is far more logical, just, and humane to treat the irreversible loss of higher brain functions as legal death, reserving the right of the patient or the patient's family to maintain biological existence. Including neocortical death within the law's approach to death ensures that the consequences and rights that flow from whole brain and cardiopulmonary death will also attend neocortical death.

Distinguishing legal from biological death also reverses the moral and legal dilemmas in "tragic choice" cases in profound ways. When attending physicians determine that a person has irreversibly lost all those cerebral qualities which distinguish human life, all artificial life-support systems and nourishment may be terminated because patients in such a state are legally dead. Relatives, guardians, and physicians will not be forced to obtain judicial approval to terminate nourishment or cardiorespiratory support regimens, nor will families be motivated to end the biological existence of a family member out of a desire to achieve the financial and legal results that follow a determination of death.

Giving legal effect to neocortical death will simplify and purify the decision to end the biological existence of a family member and assure that the right results are reached for the right reasons. In cases involving neocortical brain death, guardians, families, and courts would no longer be required to engage in a will-o'-the-wisp endeavor to discover what non-cognitive patients thought about death, or would have thought about death even if they had not thought about it or had never been

capable of thinking at all. No longer will the law favor artificially
maintaining someone in an inhuman state.

T. S. Eliot captured the essential point in the epigraph from Petro-
nius' *Satyricon*, which appears at the beginning of *The Wasteland*. After
granting the Sibyl of Cumae the gift of eternal life, Apollo does not
grant her eternal youth, and consequently her body shrivels up until she
lives in a bottle:

For I saw with my own eyes the Sibyl hanging in a jar at Cumae, and when the acolytes
said, 'Sibyl, what do you wish?' she replied, 'I wish to die' ([63], p. 459).

Vanderbilt University
Nashville, Tennessee

NOTES

[1] St. Augustine's next sentence provides a response: "The answer is: every kind of death"
([100], p. 277).
[2] Cardiopulmonary cessation or the irreversible loss of vital fluid flow (air and blood)
historically has served as the criterion for determination of death [94], ([72], p. 689).
[3] Neocortical death defines a clinical condition that reflects destruction of the critical
elements of the central nervous system, leaving the patient in an irreversible non-cognitive
condition. The terms "persistent vegetative state," "non-cognitive state," "apallic syndrome," "cerebral death" (irreversible destruction of both cerebral hemispheres), and
"irreversible lesions of the mid-brain reticular formation," would be embraced within use
of this term when diagnosis confirmed the irreversible loss of consciousness and cognitive
function ([70], pp. 184–185; [7], pp. 6–10; [76], p. 632–635). The state of residual sentience
known as the "lock-in syndrome," in which patients suffer paralysis of all four extremities
and the lower cranial nerve yet retain consciousness, is not included within the definition
of neocortical death.
[4] Ala. Code secs. 22–31–1 (1974); Alaska Stat. sec. 09.65.120 (1983); Ark. Stat. Ann.
secs. 82–537 to 82–538 (1976 & supp. 1985); Cal. Health & Safety Code, secs. 7180–7183
(1970 & supp. 1986); Colo. Rev. Stat. sec. 12–36–136 (Supp. 1985); Conn. Gen. Stat.
Ann., sec. 19a–278(b) & (c) (1984 Special Pamphlet) (part of Uniform Anatomical Gift
Act, 8A U.L.A. 15 (1975)); D.C. Code Ann., sec. 6–2401 (1982); Fla. Stat., sec. 382.085
(Supp. Pamphlet 1974–1983); Ga. Code Ann., sec. 88–1716 (Supp. 1984); Hawaii Rev.
Stat., sec. 327 C–1 (1978, am. 1979 and 1982); Idaho Code, sec. 54–1819 (1981); Ill. Ann.
Stat. ch. 110 1/2, sec. 302 (1978) (part of Uniform Anatomical Gift Act, 8A U.L.A. 15
(1975)); Iowa Code Ann., sec. 702.8 (1976); Kan. Stat. Ann, sec. 77–202 (1983 Supp.);
La. Rev. Stat. Ann., sec. 9:111 (1985 Supp.); Me. Rev. Stat Ann. tit. 22, secs. 2811 to
2813 (1985 Supp.); Md. Ann. Code Health-Gen., secs. 5–201 to 202 (1984 Supp.); Mich.
Stat. Ann, sec. 14.15 (1021 to 1023) (Callaghan 1979); Miss. Code Ann., secs. 41–36–3
(1981); Nev. Rev. Stat., sec. 451.007 (1979); N.M. Stat. Ann., sec. 12–2–4 (1978); N.C.
Gen. Stat., sec. 90–323 (1979); Ohio Rev. Code Ann., sec. 2108.30 (Baldwin 1983

Supp.); Okla. Stat. Ann. tit. 63, sec. 1–301(g) (1975); Ore. Rev. Stat., sec. 146.001 (1983 Replacement Part); Penn. Stat. Ann. tit. 35, secs. 10201 to 10203 (1984 Supp.); R.I. Gen Laws, sec. 23–4–16 (1982 Supp.); Tenn. Code Ann., sec. 68–3–501 (1982); Tex. Rev. Civ. Stat. Ann. art. 4447t (1985 Supp.); Vt. Stat. Ann. tit. 18, sec. 5218 (1985 Supp.); Va. Code, sec. 54–325.7 (1973, am. 1979); W.Va. Code, Sec. 16–10–1 to 16–10–3 1984 Supp.) Wis. Stat., sec. 146.71 (1984 Supp.); Wyo. Stat., sec. 35–19–101 (1985 Supp.).

[5] Colorado legislatively adopted a whole-brain death definition, Colo. Rev. Stat., sec. 12–36–36 (Supp. 1985), after the Colorado Supreme Court judicially recognized whole-brain death in *Lovato* v. *District Court* [27]. Illinois adopted whole-brain death in *In re Haymer* [15], a case involving termination of a ventilator system from a whole-brain dead patient. The court took the view that the whole-brain death definition set out in Illinois' version of the Uniform Anatomical Gift Act was limited to that particular statute ([15], p. 943).

[6] Statutes in Arkansas, Connecticut (as part of its formulation of the Uniform Anatomical Gift Act), Nevada, North Carolina, Oklahoma, and West Virginia take this approach (*see supra*, note 4). The Uniform Brain Death Act (sec. 1, 12 U.L.A. 17 [Supp. 1985]) recommended in 1978 by the National Conference of Commissioners on Uniform State Laws also employs a singular whole-brain focus: "For legal and medical purposes, an individual who has sustained irreversible cessation of all functioning of the brain, including the brain stem, is dead. A determination under this section must be made in accordance with reasonable medical standards." Similarly, the Law Reform Commission of Canada proposed a statute that makes the irreversible cessation of all brain functions of the entire brain the sole definition of death [77].

[7] Statutes of this type are in force in California, Colorado, the District of Columbia, Georgia, Idaho, Kansas, Maine, Maryland, Mississippi, Montana, New Mexico, Ohio, Oregon, Pennsylvania, Rhode Island, Tennessee, Vermont, Virginia, and Wisconsin. States that have judicially recognized an alternative definition based on either whole-brain or cardiopulmonary criteria are: Arizona, Illinois, Indiana, Nebraska, New Jersey, Massachusetts, and Washington.

[8] The Uniform Determination of Death Act, 12 U.L.A. sec. 1, 271 (Supp. 1985), provides: "An individual who has sustained either (1) irreversible cessation of circulatory and respiratory functions, or (2) irreversible cessation of all functions of the entire brain, including the brain stem, is dead. A determination of death must be made in accordance with accepted medical standards."

[9] Alabama, Alaska, Florida, Hawaii, Iowa, Louisiana, New York, and Texas apply this formulation, first advanced by Capron and Kass [54].

[10] The Law and Medicine Committee of the American Bar Association notes other advantages: it ". . . permits judicial determination of the ultimate facts of death; permits medical determination of the evidentiary fact of death; avoids religious determination of any facts; avoids prescribing the medical criteria; enhances changing medical criteria; enhances local medicine practice tests; covers the three known tests ('brain, beat, and breath'); covers death as a process (medical preference); covers death as a point in time (legal preference); avoids passive euthanasia; avoids active euthanasia; covers current American and European medical practices; covers both civil and criminal law; covers current American judicial decisions; avoids nonphysical sciences" ([109], p. 430).

[11] As one court observed: "Now, however, we are on the threshold of new terrain – the

penumbra where death begins but life, in some form, continues. We have been led to it by the medical miracles which now compel us to distinguish between 'death,' as we have known it, and death in which the body lives in some fashion but the brain (or a significant part of it) does not" ([3], p. 1344).

Dr. Earl A. Walker, a renowned neurosurgeon, notes: "After the concept of brain death has been introduced and generally accepted, the question was raised in philosophical and medical discussions as to whether the lack of function of an essential part of the central nervous system might not be considered as death. . . These philosophical issues may become the medical problems of the future. Obviously the concept of brain death has opened a Pandora's box that will trouble mankind for a long time" ([111], p. 261), [103, 112]).

[12] Dougherty [59] notes that the dissemination of training in cardiopulmonary resuscitation may be increasing the incidence of the persistent vegetative state at the expense of whole brain death and cardiopulmonary death (see also [92], pp. 283-284). An estimated 10,000 patients in the United States are neocortically dead but are being kept biologically alive by artificial means [68, 80].

[13] "The longest recorded coma was that undergone by Elaine Esposito of Tarpon Springs, Florida. She never stirred after an appendectomy on August 6, 1941, when she was 6, in Chicago. She died November 24, 1978, age 43 years 357 days, having been in a coma for 37 years 111 days" ([81], p. 37).

[14] In 1975, Karen Ann Quinlan lapsed into an irreversible persistent vegetative state. On June 11, 1985, Ms. Quinlan was pronounced dead after cessation of heart and lung functions [90].

[15] "The million-dollar courtroom drama had finally closed. The jetsetting Dane, the raven-haired mistress, the German-born maid, the vengeful stepchildren, had taken their curtain calls. Life had changed for all the cast members of the von Bülow play except for one: Sunny. She spent the day after her husband's acquittal like all others in the half-life of irreversible coma. She lay in a bed behind the guarded door of the $725-a-day room in Columbia Presbyterian Hospital in New York City. For the 1,632nd day she did not see anything or hear anything or feel anything or taste anything. The physical therapist came in to exercise her limbs and to turn her from one side to the other to prevent bedsores. Her hair was combed, makeup applied, teeth brushed. Her 80-pound body was fed through a tube" ([68], p. 27).

[16] In Aristotle's terms: "Now, in the case of animals, life is defined by their capacity for sense perception, and in the case of man by the capacity for sense perception and for thought. But a capacity is traced back to its corresponding activity, and it is activity that counts. Consequently, life in the true sense is perceiving or thinking" ([44], p. 265). David Rabin, a victim of amyotrophic lateral sclerosis (ALS), and his wife, Pauline Rabin, urge that patients who are victims of diseases such as ALS "can take heart in the motto *cogito ergo sum* . . . " ([96], p. 52).

[17] See *Eichner* v. *Dillon* [4], *Foody* v. *Manchester Mem. Hosp.* [5], *In re Colyer* [9], *In re Conroy* [11], *In re P.V.W.* [19], *In re Spring* [21], *In re Storar* [22], *In re Torres* [23], *John F. Kennedy Mem. Hosp.* v. *Bludworth* [25], *Leach* v. *Akron Gen. Med. Ctr.* [26], *Severns* v. *Wilmington Med. Ctr. Inc.* [33], *Superintendent of Belchertown State School* v. *Saikewicz* [38].

[18] In *Superintendent of Belchertown State School* v. *Saikewicz* [38], the court approved the guardian's decision to withhold chemotherapy from a 67-year-old incompetent patient

with an I.Q. of 10 and a mental age of approximately two years and eight months, who suffered from acute leukemia. In *In re Spring* [21], the family was allowed to terminate life-sustaining hemodialysis treatment being administered to a 78-year-old patient, who although conscious and capable of limited cognitive functions, suffered from senility. In *In re Conroy* [11], the removal of a feeding tube from an 83-year-old nursing home patient with severe physical and mental impairments was approved. In *In re Hier* [16], the probate judge's determination of substituted judgment that a seriously ill 92-year-old incompetent person with a history of mental illness would, if competent, decline surgical procedures necessary for reinsertion of feeding tubes was upheld.

[19] Patients with no higher brain functions have lost the abilities to think, feel, listen, or communicate. Patients who, although severely demented, terminally ill, or critically handicapped, nevertheless retain consciousness, have the capacity, although perhaps very limited, to think, feel, listen, or communicate. A person with dementia or an unborn fetus possesses the potential capacity to think and, therefore, is not dead. St. Thomas Aquinas noted the importance of potentiality. "Every kind of being is divided into potentiality and act . . . Now, it is noticeable that whatever has a soul is not always actual in the sense of vitally acting; so in the soul's definition it is said that it is *the act of a body having life potentiality* . . . " ([41], pp. 215–216).

Some critics have suggested that treating irreversibly non-cognitive patients as dead is but one step short of justifying active euthanasia or abortion ([46], pp. 340–341; [94], p. 40). This argument, however, is misleading. Rather than undermine the personhood status of the senile, severely sick, or unborn, a neocortical approach to death arguably strengthens the case for human rights in these settings by stressing consciousness and the capacity for thinking – not privacy, the relative quality of life, or the substituted judgment of others – as the essential test for dispensing or defining death. By contrast, alternative formulations that focus on other values such as privacy, quality of life, or the perceived best interests of the patient, do not prevent patients (or the unborn) from being put to death even though there may be the presence or potential for cognitive functions. This does not mean, however, that recognizing neocortical death would rule out terminating or withdrawing care from patients who retain limited cognitive function. With legal death properly defined to include neocortical death, medical decision-makers can directly face the distinct issue of withholding or terminating life-supports from patients who are alive and sapient but afflicted by terminal illness, chronic pain, extremely diminished abilities, or a limited life-expectancy.

[20] The court phrased the issue: "We deal here with the physician's responsibility in a case of a patient who, though not 'brain dead,' faces an indefinite vegetative existence without any of the higher cognitive functions" ([1], pp. 1013–1014).

[21] As the Washington Supreme Court later summarized the holding in *Colyer*, "The guardian of a person in a chronic persistent vegetative state can consent to the withdrawal of life support systems, at least where the family, the treating physicians, and a physicians' prognosis board agree as to proper treatment' ([1], p. 1367).

[22] The court concluded: "If the incompetent patient's immediate family, after consultation with the treating physician and the prognosis committee, all agree with the conclusion that the patient's best interests would be advanced by withdrawal of life sustaining treatment, the family may assert the personal right to refuse life sustaining treatment without seeking prior appointment of a guardian" ([14], p. 1372).

[23] *Foody* v. *Manchester Mem. Hosp.* [5] granted the family of an irreversibly non-

cognitive patient injunctive relief to restrain hospital personnel and physicians from
continuing the use of artificial devices to maintain cardiopulmonary functions. *Severns* v.
Wilmington Med. Ctr., Inc. [33] granted the husband of an irreversibly unconscious wife
guardianship for the purpose of removing life-sustaining machinery. *John F. Kennedy
Mem. Hosp.* v. *Bludworth* [25] permitted consenting family members to terminate
life-sustaining procedures without the necessity of appointment of a legal guardian where
the patient is irreversibly comatose and non-cognitive. *In re P.V.W.* [19] authorized the
parents of a severely brain-damaged, irreversibly comatose, respirator-dependent newborn to have life-supports discontinued. *In re Dinnerstein* [12] validated the order not to
resuscitate an irreversibly non-cognitive patient in a vegetative state with Alzheimer's
disease. *In re Torres* [23] authorized the conservator to have life-sustaining respirator
treatment terminated in the case of an irreversibly non-cognitive and unconscious adult
patient. *In re Storar* [22] granted the guardian of an 83-year-old, non-cognitive patient in a
chronic, persistent vegetative state the authority to have the respirator discontinued.
Leach v. *Akron Gen. Med. Ctr.* [26] granted the guardian authority to discontinue all
life-support measures for a terminally ill adult in a permanent vegetative state.

[24] *See* N.C. Gen. Stat., sec. 90–322 (Supp. 1983); Ore. Rev. Stat., sec. 97.083 (Supp.
1983); and Va. Code, sec. 54–3258: 6 (Supp. 1985).

[25] Dr. Ronald Cranford, Chairman of the Ethics and Humanities Committee of the
American Academy of Neurology, examined Mr. Brophy and testified: "He is in a
permanent vegetative state. There's no possibility he will recover. He is no more capable
of experiencing pain than someone who is in a coma, brain-dead, or dead" ([79], p. 25).

[26] While the New Jersey Supreme Court eliminated any distinction between nourishment
and other types of life-support systems in *In re Conroy*, the patient, Claire Conroy, was
described by the Court as "an elderly nursing-home resident who is suffering from serious
and permanent mental and physical impairments, who will probably die within one year
even with treatment" ([11], p. 1219).

[27] Kamisar argues that "we cannot enter the minds of comatose people to learn if they
wish to struggle on. But we can end the fiction of presuming to speak on their behalf.
Instead, let courts be honest and say life-support systems should be turned off not because of
patient's wishes but, alas, because they think the patient is 'better off dead'" ([73], p. 19).

[28] This example has its real-life parallel in the Claus von Bülow trial, which resulted in an
acquittal of the defendant. Instead of being indicted for the attempted murder of his wife,
Mr. von Bülow could have been charged with first-degree murder.

[29] The President's Commission seemed to think so: "[T]he implication of the personhood
and personal identity arguments is that Karen Quinlan, who retains brain stem function
and breathes spontaneously, is just as dead as a corpse in the traditional sense. The
Commission rejects this conclusion and the further implication that such patients could be
buried or otherwise treated as dead persons" ([94], p. 40). And Bernat, Culver and Gert
also urge that "a practical problem also arises in considering chronically vegetative
patients with spontaneous ventilation to be dead. To bury such patients while they breathe
and have a heartbeat, most would view as at least ethically unacceptable" ([47], p. 391).

[30] Dr. William Schwartz, of Tufts University, has noted: "If I know that keeping someone
alive for another few months at a cost of $100,000 will mean that sum is not available for
the care of several people with severe hip or heart disease, then I have a new kind of moral
dilemma: Is it proper to use that limited resource for maintaining life of poor quality for a
few months, denying care to others for whom I also have a responsibility?" ([98], p. 25).

The cost of biologically maintaining a neocortically dead body is substantial, as Goodman points out: "[P]atients . . . in a chronic vegetative state . . . are in hospital beds at an average cost of $150,000 per person per year" ([68], p. 27), and Mancusi pointed out that the care for Mr. Brophy costs $10,500 per month ([101], p. 25).

[31] *See supra*, note 23.

[32] "Today we are witnessing the birth of a new technology for the study of the human brain, based on the use of radioactive tracer molecules carbon-11, fluorine-18, and oxygen-15. . . . After injection, the biodistribution of the tracers is portrayed by positron-emission tomography (PET). In a typical study, the cyclotron-produced radioactive atom is incorporated into a substrate, such as glucose or fatty acid, or into a drug . . . The intravenously injected glucose or oxygen is metabolized in bioenergetic pathways . . . Serial images of regions within the brain are produced by the measurement of gamma rays coming from within different regions of the brain. A PET scanner is similar to a computed tomography (CT) scanner except that gamma rays are emitted from within the patient instead of traveling across the brain as in x-ray CT. As a result of work performed in several countries beginning in the mid-1970s, it is now well established that measurable increases in regional blood flow and in glucose and oxygen metabolism accompany mental functions, including perception, cognition, and emotion. . . . The ability to study neurotransmitters and neuroceptors as well as the substrate metabolism of the brain makes it increasingly likely that every major university medical center will have a cyclotron and positron tomographic device within the next 5 to 10 years. It is predictable that regional cyclotrons, whether operating at university medical centers or commercially developed, will soon provide short-lived tracers labeled with carbon-11 or flouride-18 to community hospitals throughout a city . . . Another developing clinical application is the measurement of regional oxygen metabolism as an indicator of the survivability of involved regions of the brain in patients with strokes . . . Perhaps it is not overstating the case to say that in positron-emitting tracers in community hospitals and PET in major medical centers, we now have a set of eyes that permits us to examine the chemistry of the mind" ([110], pp. 45–46), [62].

[33] Leenders *et al.* note: "Positron Emission Tomography (PET) has evolved in recent years into a powerful clinical research technique for the study of the physiology and pathophysiology of the human brain *in vivo*. Using specific tracers labelled with positron emitting isotopes and a tomographic scanning technique . . . it is possible to measure quantitatively in an essentially non-invasive way local tissue functions such as cerebral blood flow, oxygen and glucose utilisation" ([78], p. 1; [62], p. 844).

[34] For transplants to be successful, a viable, intact organ is needed. Unlike whole brain dead donors, neocortically dead donors would not be subject to the precarious time clock of a few scant hours or days in which to transplant a kidney, heart, or lung, for example. The appointment by the Secretary of Health and Human Services of a panel to advise the Secretary on improving the nation's organ transplant network pursuant to congressional legislation to study and fund donor transplant programs evidences the important need to improve the transplant system [43], [84], [89].

[35] Uniform Anatomical Gift Act, sec. 3, 8A U.S.A. 41 (1975).

[36] Some commentators question the ethics of such a practice ([71], p. 37; [113], p. 76.)

[37] An attempted murder charge might be sought if the defendant believed the body were alive at the time the defendant acted [97].

[38] For example, in an Illinois case, a father slammed a severely deformed **infant onto the**

delivery room floor twenty-nine minutes after birth, killing it. The father was charged with murder [87]. Assuming the child was born with severe brain damage to the point of being irreversibly unconscious and non-cognitive (a fact not capable of proof in this case, given the short time period before the infant was killed), it could be contended the father committed no crime since, under a neocortical death definition, the infant was born legally dead although biologically alive. However, under a neocortical definition, or under the right of privacy right-to-die cases, the parents could have effected termination of biologic existence if they had so desired [18].

Unlike disconnecting life-support measures or nourishment systems, and unlike active termination by chemical injection, however, the father's acts in this case may well call for criminal sanctions, even assuming his spouse and family would have sought withdrawal of treatment and cessation of food and water. For example, a section of the Texas Penal Code, entitled "Abuse of Corpse," provides in pertinent part: "A person commits an offense if, not authorized by law, he intentionally or knowingly . . . treats in a seriously offensive manner a human corpse" [Tex. Penal Code Ann., sec. 42.10(a)(1) (Vernon 1974)].

In a Miami, Florida, case, a 25-year-old father shot and killed his 3-year-old daughter who was irreversibly unconscious and non-cognitive as a result of a freakish accident involving a reclining chair[85].

BIBLIOGRAPHY

1. *Barber* v. *Superior Court*, 147 Cal. App.3d 1006, 195 Cal. Rptr. 484 (1983).
2. *Commonwealth* v. *Golston*, 373 Mass. 249 N.E.2d 744 (1977), *cert. denied*, 434 U.S. 1039 (1978.)
3. *Douglas* v. *Sw. Life Ins. Co.*, 374 S.W.2d 788 (Tex. Civ. App. Div. 1964).
4. *Eichner* v. *Dillon*, 73 A.2d 431, 426 N.Y.S.2d 517 (N.Y. App. Div. 1980), *aff'd sub. nom. In re Storar*, 52 N.Y. 2d 363, 420 N.E. 2d 64, 438 N.Y.S. 2d 266, *cert. denied sub. nom., Storar* v. *Storar*, 454 U.S. 858 (1981).
5. *Foody* v. *Manchester Mem. Hosp.*, 40 Conn. Sup. 127, 482 A2d 713 (Conn. Super.Ct. 1984).
6. *Hake* v. *Manchester Township*, 98 N.J. 302, 486 A.2d 836 (1985).
7. *In re Barry*, 445 So2d 365 (Fla. Dist. Ct. Appl. 1984).
8. *In re Bowman*, 94 Wash. 2d 406, 617 P.2d 731 (1980).
9. *In re Colyer*, 99 Wash. 2d 114, 660 P.2d 738 (1983).
10. *In re Conroy*, 190 N.J. Super. 453, 464 A.2d 303 (1983).
11. *In re Conroy*, 98 N.J. 321, 486 A.2d 1209 (1985).
12. *In re Dinnerstein*, 6 Mass. App. 466, 381 N.E.2d 134 (1978).
13. *In re Estate of Schmidt*, 261 Cal. App.2d 262, 67 Cal. Rptr. 847 (1968).
14. *In re Hamlin*, 102 Wash.2d 810, 689 P.2d 1372 (1984).
15. *In re Haymer*, 115 Ill. App3d 349, 450 N.E.2d 940 (1983).
16. *In re Hier*, 18 Mass. App. 200, 464 N.E.2d 959 (1984).
17. *In re Ingram*, 102 Wash.2d 810, 689 P.2d 1363 (1984).
18. *In re L.H.R.*, 253 Ga. 439, 321 S.E.2d 716 (1984).
19. *In re P.V.W.*, 424 So.2d 1015 (La. 1982).
20. *In re Quinlan*, 70 N.J. 10,355 A.2d 647, *cert. denied*, 429 U.S. 922 (1976).

21. *In re Spring*, 380 Mass. 629, 405 N.E.2d 115 (1980).
22. *In re Storar*, 52 N.Y.2d 363, 420 N.E.2d 64, 438 N.Y.S.2d 266 (1981), *cert. denied* 454 U.S. 858 (1981).
23. *In re Torres*, 357, N.W.2d 332 (Minn. 1984).
24. *In re Welfare of Bowman*, 94 Wash2d 407, 617 P.2d 731 (1980).
25. *John F. Kennedy Mem. Hosp.* v. *Bludworth*, 452 So.2d 921 (Fla. 1984).
26. *Leach* v. *Akron Gen. Med. Ctr.*, 68 Ohio Misc. 1, 426 N.E.2d 809 (Ohio Com. Pl. 1980).
27. *Lovato* v. *District Court*, 198 Colo. 419, 601 P.2d 731 (1980).
28. *Mack* v. *City of Minneapolis*, 333 N.W.2d 744, 746 (Minn. 1983).
29. *Merkrus Estate*, 24 Pa. Fiduc. 249 (Orph. Ct. 1974).
30. *Patricia Brophy, wife of Paul Brophy* v. *New England Sinai Hosp.*, No. 85–E–0009–G–1, *Slip op.* (Probate and Family Court, Norfolk Div., Dedham, Mass. Oct. 21, 1985), *rev'd*, No. N–4152, Mass. Supreme Judicial Ct., Sept. 11, 1986.
31. *People* v. *Eulo*, 63 N.Y.2d 341, 472 N.E.2d 286, 482 N.Y.S.2d 436 (1984).
32. *Schmitt* v. *Pierce*, 344 S.W.2d 120 (Mo. 1961).
33. *Severns* v. *Wilmington Med. Ctr., Inc.*, 421 A2d 1334 (Del. 1980).
34. *Smith* v. *Smith*, 229 Ark. 479, 317 S.W.2d 275 (1958).
35. *State* v. *Fierro*, 124 Ariz 182, 603 (1979).
36. *State* v. *Meints*, 212 Neb 410, 322 N.W.2d 809 (1982).
37. *State* v. *Watson*, 191 N.J. Super. 464, A.2d 590 (1983).
38. *Superintendent of Belchertown State School* v. *Saikewicz*, 373 Mass 728, 370 N.E.2d 4178, 823–25 (1977).
39. *Swafford* v. *State*, 421 N.E.2d 596, 602 (Ind. 1981).

REFERENCES

40. Ad Hoc Committee of the Harvard Medical School to Examine the Definition of Brain Death: 1968, 'A Definition of Irreversible Coma', *Journal of the American Medical Association* **205**, 337.
41. Aquinas, St. T.: 1972, *Summa Theologiae* (ed.), *An Aquinas Reader*, Image Books, A Div. of Doubleday and Company, Inc., Garden City, N.Y.
42. *American Medical News*: 1985, 'Former Senator Pleads for Dignified Death' (Oct. 25), 13.
43. *American Medical News*: 1985 'Legislation to Spur Organ Donations Urged' (Oct. 18), 24.
44. Aristotle: 1962, *Nicomachean Ethics*, Bobbs-Merrill Company, Inc., The Library of Liberal Arts.
45. Arnold J.: 1977, 'Neomorts', *University of Toronto Medical Journal* **54**, 35–37.
46. Beresford: 19 , 'Cognitive Death: Differential Problems and Legal Overtones', *Annals of the New York Academy of Sciences* **315**, 339–348.
47. Bernat, J., Culver, C., and Gert, B.: 1981, 'On the Definition and Criterion of Death', *Annals of Internal Medicine* **94**, 389–394.
48. Bernat, J., Culver, C., and Gert, B.: 1982, 'Defining Death in Theory and Practice', *Hastings Center Report* **12**, 5–9.

49. Black, P.: 1978, 'Brain Death', *New England Journal of Medicine*, 299, 338–344.
50. Boswell, J.: 1934, *Boswell's Life of Johnson*, Oxford University Clarendon Press, Oxford.
51. Brierly, J., Adam, J., Graham, D., and Simpson, J.: 1971, 'Neocortical Death After Cardiac Arrest', *Lancet* 2, 560–565.
52. Bronowski, J.: 1973, *The Ascent of Man*, Little, Brown, New York.
53. Capron, A.: 1978, 'Legal Definition of Death', *Annals of the New York Academy of Sciences* 315, 349–359.
54. Capron, A., and Kass, L.: 1972, 'A Statutory Definition of the Standards for Determining Human Death: An Appraisal and a Proposal', *University of Pennsylvania Law Review* 121, 87–188.
55. Carroll, L.: 1982, *Alice's Adventures in Wonderland*, Oxford, p. 14.
56. Conference of Royal Colleges and Faculties of the United Kingdom: 1976, 'Diagnosis of Brain Death', *Lancet* 2, 1069–1070.
57. Cranford, R. and Smith, H.: 1979, 'Some Critical Distinctions Between Brain Death and Persistent Vegetative State', *Ethics in Science and Medicine* 6, 199–209.
58. Descartes, R.: 1910, *Discourse on Method*, Harvard, Cambridge, Mass.
59. Dougherty, J., Rawlinson, T., Levy, D., and Plum, F.: 1981, 'Hypoxialschemic Brain Injury and the Vegetative State: Clinical and Neuropathic Correlation', *Neurology* 31, 991–997.
60. Dworkin, R.: 1973, 'Death in Context', *Indiana Law Journal* 48, 623–639.
61. Dukeminer, J. and Johanson, S.: 1978, *Family Wealth Transactions, Wills, Trusts, and Estates*, Little Brown & Co., New York.
62. Editorial: 1984, 'Pet Scans Relate Clinical Picture to More Specific Nerve Function', *Journal of the American Medical Association* 253, 943–949.
63. Eliot, T.S.: 1973, 'The Wasteland', in R. Ellman and R. O'Clair (eds.), *The Norton Anthology of Modern Poetry*, W.W. Norton & Co., New York.
64. Engelhardt, H.T., Jr.: 1975, 'Defining Death: A Philosophical Problem for Medicine and Law', *American Review of Respiratory Disease* 112, 587–590.
65. Engelhardt, H.T., Jr.: 1982, 'Medicine and the Concept of Person', in T. Beauchamp and W. Walters (eds.), *Contemporary Issues in Bioethics*, Wadsworth Pub. Co., pp. 94–112.
66. Fletcher, J.: 1974, 'New Definitions of Death', *Prison* 2, 13– . 2 *Am. J. Law & Med.* p. 1(76); 2 *Annual Review of Soc.* 2/231 (76).
67. Gert, B.: 1971, 'Personal Identity and the Body', *Dialogue* 10, 458–478.
68. Goodman, E.: 1985, 'When the Choice is Irreversible Coma or Turning off the Feeding Tube', *Boston Globe* (June 13), 27.
69. Green, M. and Wikler, D.: 1980, 'Brain Death and Personal Identity', *Philosophy and Public Affairs* 9, 105–133.
70. Ingvar, D.H., Brun, L., Johnson, L., and Samuelson, S.M.: 1978, 'Survival After Severe Cerebral Anoxia, With Destruction of the Cerebral Cortex: The Apallic Syndrome', *Annals of the New York Academy of Sciences* 315, 184–214.
71. Jonas, H.: 1974, 'Against the Stream: Comments on the Definition and Redefinition of Death', in H. Jonas, *Philosophical Essays*, University of Chicago Press, Chicago, pp. 132–140.

72. Joynt, A.: 1984' 'A New Look at Death', *Journal of the American Medical Association* **252**, 680–682.
73. Kamisar, Y.: 1985, 'The Real Quinlan Issue', *The New York Times* (June 13), 19.
74. Korein, J.: 1978, 'Terminology, Definitions, and Usage', *Annals of the New York Academy of Sciences* **315**, 6–10.
75. Korein, J.: 1978, 'The Problem of Brain Death: Development and History', *Annals of the New York Academy of Sciences* **315**, 19–38.
76. Ladanyi, K.: 1984, 'Residual Sentience and Cognitive Death: Ethical Issues in Brain Death and Persistent Vegetative State', *Canadian Medical Association Journal* **131**, 632–636.
77. Law Reform Commission of Canada: 1981, *Report on the Criteria for the Determination of Death*, Minister of Supply and Services Canada, Ottawa.
78. Leenders, P., Gibbs, F., Franckowiak, J., Lammertsma, T., and Jones, A.: 1984, 'Positron Emission Tomography of the Brain: New Possibilities for the Investigation of Human Cerebral Pathophysiology', *Progress in Neurobiology* **23**, 1–38.
79. Mancusi, P.: 1985, 'Wife Asks Feeding Tube to be Removed', *Boston Globe* (May 9), 25.
80. Mancusi, P.: 1985, 'Karen Ann Quinlan's Legacy: Questions on Dying That Have Yet to be Answered', *Boston Globe* (June 13), 3.
81. McWhirter, N. (ed.): 1984, *The Guiness Book of World Records*, Bantam Books, New York.
82. Moraczewski, A. and Shlowlater, J.: 1982, *Determination of Death: Medical, Ethical and Legal Issues*, Cath. Health Assoc., New York.
83. National Institutes of Health: 1976, 'An Appraisal of the Criteria of Cerebral Death: A Summary Statement', *Journal of American Medical Association* **237**, 982–986.
84. *New York Times*: 1985, 'A Donor Shortage Cited by Surgeon' (Jan. 24), 11.
85. *New York Times*: 1985, 'Comatose Toddler Shot to Death; Heartsick Father Turns Himself In' (June 30), 15.
86. *New York Times*: 1985, 'Karen Ann Quinlan, 31, Dies: Focus of '76 Right to Die Case' (June 12), A1.
87. *New York Times*: 1985, 'Mistrial in Killing of Malformed Baby Leaves Town Uncertain About Law' (Feb. 15), 9.
88. *New York Times*: 'Pact on Right to Die Cited in Court Case' (September 18), A10.
89. *New York Times*: 1985, 'Panel on Transplants Named' (Jan 26), A16, col. 1.
90. *New York Times*: 1985, 'Quinlan Case' (June 13), 24.
91. Olinger, F.: 1975, 'Medical Death', *Baylor Law Review* **27**, 22–26.
92. Pampliglione, F., Chaloner, T., and O'Brien, P.: 1978, 'Transitory Ischemia/Anoxia in Young Children and the Prediction of Quality of Survival', *Annals of the New York Academy of Sciences* **315**, 281–291.
93. Pascal, B.: 1965, *Pensees*, Librairε Larousse, Paris, pp. 74, 117.
94. President's Commission for the Study of Ethical Problems in Medicine and Biomedical and Behavioral Research: 1981, *Defining Death: A Report on the Medical, Legal, and Ethical Issues in the Determination of Death*, U.S. Government Printing Office, Washington, D.C.
95. Puccetti, R.: 1976, 'The Conquest of Death', *The Monist* **59**, 259 ff.

96. Rabin, D. and Rabin, P.: 1985, 'Credo for Creeping Paralysis: Cogito Ergo Sum', in D. Rabin and P. Rabin (eds.), *To Provide Safe Passage: The Humanistic Aspects of Medicine*, Philosophical Library, New York, pp. 48–53.

97. Rosenfeld, .: 1970, 'The Heart, the Head, and the Halakhah', *New York State Journal of Medicine* (Oct. 15), 2615–2619.

98. Schwartz, W.: 1984, 'U.S. Medicine Can't Do Everything for Everybody', *U.S. News and World Report* (June 25), 25.

99. Smith, J. and Hogan, B.: 1983 (5th ed.), *Criminal Law*, Butlerworth, London.

100. St. Augustine: 1958, *The City of God*, Image Books.

101. Thomas, D.: 1973, 'A Refusal to Mourn the Death, by Fire, of a Child in London', in R. Ellman and R. O'Clair (eds.), *the Norton Anthology of Modern Poetry*, W.W. Norton & Co., New York.

102. Thornton, ., and Staff: 1983 'Death and the Life Insurance Policy: What Hath Modern Medicine Wrought?', *Oklahoma Law Review* **36**, 285–308.

103. Tomlinson, T.: 1984, 'The Conservative Use of the Brain-Death Criterion: A Critique', *Journal of Medicine and Philosophy* **9**, 377–393.

104. Tyler, and Robertson, : 1984, 'Impact of the Brain Death Ruling in Washington State', *Western Journal of Medicine* **140**, 625–627.

105. Ufford, W.: 1980, 'Brain Death/Termination of Heroic Efforts to Save Life: Who Decides?', *Washburn Law Journal* **19**, 225–259.

106. Veatch, R.M.: 1975, 'The Whole-Brain-Oriented Concept of Death: An Outmoded Philosophical Formulation', *Journal of Thanatology* **3**, 13–30.

107. Veatch, R.M.: 1976, *Death, Dying and the Biological Revolution: Our Last Quest for Responsibility*, Yale University Press, New Haven.

108. Veatch, R.M.: 1978, 'The Definition of Death: Ethical, Philosophical, and Policy Confusion', *Annals of the New York Academy of Sciences* **315**, 307–321.

109. Vieth, F.J.: 1978, 'Brain Death and Organ Transplantation', *Annals of the New York Academy of Sciences* **315**, 417–433.

110. Wagner, H.: 1985, 'Probing the Chemistry of the Mind', *New England Journal of Medicine* **312**, 44–46.

111. Walker, F.: 1983, 'Current Concepts of Brain Death', *Journal of Neurosurgical Nursing* **15**, 261–264.

112. Walker, F.: 1984, 'Dead or Alive?', *Journal of Nervous and Mental Disease* **172**, 639–641.

113. Walton, D.: 1980, *Brain Death: Ethical Considerations*, Purdue University Press, West Lafayette, IN.

114. Walton, D.: 1981, 'Epistemology of Brain Death Determination', *Metamedicine* **2**, 259/ff.

115. Youngner, S.J. and Bartlett, E.T.: 1983, 'Human Death and High Technology: The Failure of Whole-Brain Formulations', *Annals of Internal Medicine* **99**, 252–

PART III

THE PRESIDENT'S COMMISSION AND BEYOND

ALEXANDER MORGAN CAPRON

THE REPORT OF THE PRESIDENT'S COMMISSION ON THE UNIFORM DETERMINATION OF DEATH ACT

I have been asked to describe and assess the work of the President's Commission for the Study of Ethical Problems in Medicine and Biomedical and Behavioral Research on the problem of "defining" death. In the midst of a conference devoted to very thoughtful philosophical analyses by physicians and historians as well as philosophers, a presentation on a governmental report takes on the appearance of a thorn among the roses. Although the work of the President's Commission on this subject – presented in July, 1981, in the report *Defining Death* – had among its objectives that of being conceptually sound, it was by no means primarily a philosophical document, but instead part of the process of formulating public policy. While I believe it is fair to ask questions about the report's theoretical aspects, I also believe that any judgment on the success of this endeavor has to be made primarily in terms of its impact with lawmakers as well as its practical usefulness to physicians and others who actually have to make death determinations.

Accordingly, in this paper I will describe what the Commission took to be its objectives and why, and then set forth and analyze our results, including an examination of the major points that remain at issue.

I. TASKS AND OBJECTIVES

The Commission faced two major tasks in responding to its congressional mandate to study "the ethical and legal implications of the matter of defining death, including the advisability of developing a uniform definition of death" [27]. First, the Commission had to assess the need for a statutory "definition" of death arising from developments in medical science and practice and from the existing legal reactions to these developments. Second, the Commission had to determine the particular role that it could best play in responding to this need. This led the Commission to focus on the development of a single, uniform proposal to reformulate the publicly established standards for determining that death has occurred; the Commission also facilitated an updating of the medical criteria for diagnosing death, in order to

147

Richard M. Zaner (ed.), Death: Beyond Whole-Brain Criteria, pp. 147–169.
© 1988 *by Kluwer Academic Publishers.*

provide the "accepted medical standards" needed as part of any probable statutory response.

The central bases for an evaluation of the Commission's success in these tasks are, first, the report the Commission issued on this subject in July, 1981, which sets forth the proposed Uniform Determination of Death Act (UDDA) in the context of an 84-page report (with accompanying appendices) [26], and, second, a statement by 56 medical consultants (who included virtually every leading figure in the medical community on the subject of determination of death) that distills current, accepted medical practice in the United States for determining the occurrence of death [23].

The consultants' "Guidelines" is an important document. Indeed it was hailed as a landmark in a *JAMA* editorial when it was issued [4]. It has been widely published in the medical and nursing literature, and, I am told, has been influential in providing a single and generally accepted statement of the methods and procedures for determining death, updating the 1968 report of the Harvard Ad Hoc Committee [1]. (If nothing else, the process of forging this new consensus is an important chapter in the annals of medical diplomacy.) Nonetheless, the "Guidelines" is not, in a strict sense, the work of the President's Commission; indeed, the consultants explicitly stated that their criteria "had not been passed on by the Commission and are not intended as matters for governmental review or adoption" ([23], p. 2184). Thus, it seems appropriate, in measuring the Commission's contribution to the subject, to focus on the proposed statute and the report in which it is explicated and its probable effects are analyzed, while recognizing that the Commission also contributed to resolution of problems in this field by facilitating the production of its medical consultants' report.

In evaluating the Commission's work, we should measure its success in meeting four objectives: Did it act in a way that was medically accurate, conceptually sound, accessible to the public, and free from undue complexity? These seem to me to be generally appropriate standards for public policymaking in the biomedical field. Obviously, there is a potential for tension among these objectives. Indeed, I suspect that most criticism of the Commission's work comes down in the end to the critic's belief that the Commission emphasized one or another of these objectives to the (in the critic's view, regrettable) detriment of other (more important) objectives. From reviewing the literature, I have the sense that particularly in philosophical circles the latter two objectives – public accessibility and legal simplicity – are seen as less important than the first two. My own sense is that the formulation of public policy in the real world cannot ignore – or even downplay – the latter two objectives without either becoming unattractive to policymakers (particularly legislators) or being regarded as too uncertain to guide those who implement policy.

The thesis of this paper is that by choosing a statute that is uniform among people, not locked into current medical technology, and based on the "whole-brain" concept of death, while still permitting the use of cardiopulmonary measures, the Commission succeeded in balancing the relevant public policy objectives in an appropriate fashion. After reviewing the medical and legal situation at the time the Commission was working, I will analyze the major issues the Commission faced and defend its resolution of them. In sum, I believe the UDDA removes the major impediments to the adoption of uniform legislation; the remaining barriers come from what I believe are the misguided objections of certain religious groups.

II. THE SETTING OF THE COMMISSION'S STUDY

As I stated a moment ago, the Commission's first task was to assess the need for a statute on death in light of medical and legal developments in this field in which so much had occurred prior to January, 1980, when the Commissioners were sworn in and commenced their work. We may begin, then, by considering the clinical development that had occurred relevant to the diagnosis of death.

A. *Medical*

Until recently laymen were probably as comfortable with this subject as were physicians – a determination of death was based on such manifest signs as an absence of pulsation, absence of breathing, and fixed pupils. Actually, while these criteria have been used to determine death since ancient times, they have not always been universally accepted. It is not merely the macabre novels of the 19th century that were responsible for the horror over being buried alive – when cemeteries are dug up, a few exhumed skeletons are found to have clawed at coffin lids. Some people even took elaborate steps, such as coffins with escape mechanisms and speaking tubes to the world above, to avoid such a ghastly end [2, 3]. Well into the 18th century, respected physicians argued that putrefaction was the only sure sign of death. Only in the 19th century did the debate gradually dissipate as physicians became more competent and the public gained greater faith in them ([2], p. 30). Part of the confidence centered on the invention of the stethoscope in the mid-19th century, which enabled physicians to detect a heartbeat with increased sensitivity. The public took reassurance from this technological means of measuring what had long been regarded as one of the central signs in the diagnosis of death.

In the past 20 years, however, a divergence between the popular perception of death and that applied by the medical profession in certain cases has been brought about by the development and widespread use of mechanical respirators and related forms of treatment. Mechanical respirators are applied in some cases because of an interference with, or loss of function in, the respiratory centers of the medulla which leads respiration to stop, which in turn deprives the heart of needed oxygen, causing it to cease functioning. Thus, cessation of functioning in this portion of the brain stem can be responsible for the disappearance of the traditional signs of life – respiration and circulation. Likewise, the disappearance of respiration and circulation means that even a temporary cessation of brain function can become permanent, because brain cells suffer irreversible damage when they lack oxygenated blood for a few minutes. Consequently, the traditional vital signs used in diagnosing death reflected the interdependence of respiration, circulation, and brain function ([26], pp. 15–16).

That simple picture has been changed by the artificial respirator, which can compensate for the inability of the thoracic muscles to fill the lungs with air. These machines are not a perfect substitute for the missing natural functions when the cessation of breathing is caused by absent or deficient neural impulses, since the machines have difficulty in regulating blood gas levels precisely. But they are able to take over regulation of the rate and depth of "breathing," which is normally controlled by the respiratory centers in the medulla. Once the lungs are forced to work again, the heart will usually beat since its basic pumping functions are not dependent on external control, although impulses from the brain centers are needed to modulate the inherent rate and force of the heartbeat. Consequently, when artificial respiration provides adequate oxygenation and associated medical treatments regulate essential plasma components and blood pressure, an intact heart will **continue to beat** despite loss of brain functions.

The most frequent causes of the irreversible cessation of functioning in the whole brain are (a) direct trauma to the head, (b) massive spontaneous hemorrhage into the brain, and (c) anoxic damage from cardiac or respiratory arrest or severely reduced blood pressure ([12], p. 201). Those injuries which are traumatic typically cause a loss of blood flow to both the upper and lower portions of the brain because of cerebral edema. If deprived of blood flow for 10 to 15 minutes, the adult brain (including the brain stem) will – in the absence of barbiturates or hypothermia – completely cease functioning. Even in the absence of trauma, loss of functions of the brain neurons can also occur if circulation is impaired.

Patients who have suffered a loss of circulation to the brain for either traumatic or non-traumatic reasons can be categorized into three groups. First are those who have lost all brain functions. Such bodies lack not only consciousness, and cognitive, affective, and integrative functions, but also all reflexes controlled by the brain stem, such as the gag reflex and the drive to breathe (although some spinal reflexes may persist, as circulation to the spine is separate from that to the brain). In the absence of artificial respiration it would be apparent to all that such a person is dead. With mechanical support, however, respiration and circulation may be maintained for a limited period of time.[1]

Accurate methods were developed by medicine prior to 1980 to look around these artificially supported signs and determine whether death has occurred, using tests of brain functioning at the brain stem as well as the cortical level. Diagnostic methods had advanced considerably by 1980 beyond those set forth in the 1968 Harvard report, and many alternative formulations had been put forward.

The second group of patients who have suffered brain ischemia consists of those whose brain injury is less severe – victims of trauma who receive prompt medical attention and victims of cardiac arrest whose circulation is restored before damage has occurred throughout the brain. Since the cerebrum, particularly the cerebral cortex, is more easily injured by a loss of blood flow or oxygen than is the brain stem, a brief period without respiration or circulation, perhaps four minutes or so, typically damages the cerebral cortex permanently, although the more resistant brain stem will continue to function. Patients in this category usually require artificial support for only a limited period, after which they will be able to breathe on their own – but without recovering consciousness. Their condition is usually called the "persistent vegetative state," in which they will exhibit spontaneous involuntary movements such as facial grimaces but will remain at best "awake but unaware." Physicians' ability to prognosticate for these patients is less precise than for those in the first group. With medical and nursing support, including intravenous feeding and administration of antibiotics for recurrent pulmonary infections, such patients can survive for years – indeed, the longest reported survival exceeded 37 years ([22], p. 42).

Finally, the third group of patients is those for whom the damage has been less severe. Few who persist in an unconscious state beyond one month will recover fully, but there are some who do; additionally, a larger number of unconscious patients who need artificial support for only a brief period will recover fully [17, 21]. Although small, it is the latter group that in effect explains the existence of the first two. The respirator is not merely one more example of the Sorcerer's Apprentice at work in modern medicine. It is not, in other words, a device that serves solely to confound our ability to determine whether patients are dead, nor a device which, when administered to the victims of accidents and heart attacks, produces only persistent unconsciousness. Rather, it and the accompanying medical techniques are essential in some cases to provide the life-support that acutely ill or injured patients need if they are to recover.

A study undertaken by the President's Commission at four acute care hospitals in 1980 found 133 patients who were comatose and respirator-supported during a two-month period. Only 16 of these achieved a good to moderate recovery within 30 days. (Incidentally, those who achieved a good outcome were usually in a coma due to drug intoxication.) Similar results were found from a secondary analysis of computerized data on critically ill patients at three centers, each covering a one-year period ([26], pp. 89–101). Though a small minority, it is this 10 to 12 percent of all comatose, respirator-supported patients who, it seems to me, provide the ethical justification for the use of this technology.

B. *Law and Policy*

1. *The Public's Growing Interest* The existence of patients whose respiration and circulation could be artificially sustained plainly challenges the traditional public assumptions about the meaning of death. But that was not the only force pressing for a public response. The entire matter of death has undergone a striking change in the past fifteen years. This

change is much greater than merely the attention of the popular press to the theories of people such as Dr. Elizabeth Kübler-Ross, who began in the '60's to urge health professionals to recognize the psychological stages of the dying process and found herself becoming a cult figure [19]. It goes beyond the high school yearbook photo of Karen Ann Quinlan, whose face is probably familiar to more Americans than those of most elected officials or even many Hollywood stars. And it is a phenomenon greater than the attention paid from time to time to particularly spectacular applications of new medical technology, such as the artificial heart implanted in Dr. Barney Clark.

What is so notable from the viewpoint of public policy is the degree to which matters that were previously private have now been brought into the public light. Several hypotheses may explain why this has occurred. The enormously increased costs of medical care, which are largely borne through third-party payment, make such care – especially heroic efforts at the end of life – a matter of public concern, since the payments for care are largely paid from insurance or governmental coffers rather than by individuals. Second, the increased complexity of the procedures, and indeed, the increased complexity of the institutions in which they are applied, bring more people – and more bureaucratic formality – to the patient's bedside. Third, a greater tendency on the part of the medical profession to engage in self-scrutiny can also be discerned – Hamlet rather than Julius Caesar may now be found on the medical stage. Finally, it is impossible to overlook the rising concern – indeed, near panic – in the medical profession over the prospect of liability. Physicians increasingly seek refuge in various public forums, particularly the courts, or are brought there by dissatisfied patients and families.

While this volume concentrates on the contours of the public *policies* regarding the determination of death that have resulted from this process, we should also keep in mind the nature of the *process* itself and remember that a very different set of assumptions governs decisions about death and dying today than was true even 15 years ago before the drama of heart transplantation first gripped the public's imagination, before the plight of various comatose patients became regular front-page news, and before the first ingenious defense lawyer argued that a client was at most guilty of assault but not of murder because at the time his victim's organs were removed for transplantation the victim was still breathing (with the aid of a respirator) and hence was still alive.

2. *Legal Responses Regarding the "Definition"* Court cases of this sort were not the only legal response to the new medical capabilities to extend life or at least to camouflage the signs by which its passing had previously been discerned. As of 1969, no state had a statute "defining" death, and courts continued to reiterate the common law view that death occurred only upon "complete cessation of *all* vital functions . . . even if artificially maintained" ([28], p. 71). Yet in the following decade, doubts about the need for a new legal "definition" were overcome, and on account of the uncertainties that accompany judicial resolution of the issue, a consensus emerged that legislation was appropriate and necessary. When the Commission began its study in 1980, four model laws had been proposed, and some form of statutory definition had been adopted in 24 states ([26], pp. 62–67).

Most of the adoptions had occurred by the mid-1970's, however, and those states adopting statutes in 1978 and 1979 (and even in 1980, after our study had begun) tended to favor non-standard statutes, rather than one of the model bills. Indeed, the existence of the several models seemed to be an impediment to legislative action; despite the bills' similarity and common purpose, they were perceived as competitive, and legislators understandably wanted to avoid offending one sponsoring group (such as the bar or medical association) by favoring another group's proposal.

Such then, was the situation as the Commission began its work: changes in biomedical knowledge and practice created a need for public sanction of new methods of diagnosing death, yet despite increased public awareness of the issue, in a majority of states neither courts nor legislatures had responded by updating the standards for determining human death.

III. MEETING OBJECTIVES

As a practical matter, it seemed to the Commissioners that they would not be responding well to their mandate were they simply to promulgate one more proposed statute. Instead, I met with representatives of the three organizations that had issued model laws on death, the American Bar Association, the American Medical Association, and the National Conference of Commissioners on Uniform State Laws. These groups agreed on a single, new proposal, in place of their previous models – the Uniform Determination of Death Act.

An individual who has sustained either (1) irreversible cessation of circulatory and respiratory functions, or (2) irreversible cessation of all functions of the entire brain, including the brain stem, is dead. A determination of death must be made in accordance with accepted medical standards ([26], p. 2)

How well does this statute meet the medical, conceptual, legal, and policy objectives set forth earlier?

A. *Medical Accuracy*

First, to be medically accurate, a statute would need not only to avoid embodying scientific mistakes but also to permit good medical practices. The UDDA does both. The two standards for determining death that the statute recognizes are accepted by biomedical science as equivalent bases for diagnosing the same phenomenon. The circulatory and respiratory standard is familiar to lay people as well as to physicians; a number of studies have shown that the brain-based standard is at least as accurate a ground for declaring death ([26], pp. 24–29, 83).

In addition to American studies, British investigations also led the Conference of Royal Colleges and Faculties to conclude that death has been established when "all functions of the brain have permanently and irreversibly ceased" [11]. It is important to emphasize – given the attention sometimes paid in lay accounts to the use of an electroencephalograph in diagnosing death – that "functions" are not to be confused with "mere *activity* in cells or group of cells if such activity (metabolic, electrical, etc.) is not manifested in some way that has significance for the organism as a whole" ([26], p. 28).

The central way in which the statute permits good medical practices is by not restricting physicians to current methods or techniques. In other words, it recognizes that what is "accepted" by physicians will evolve with new scientific findings and technical advances. The law does not prelude changes in the criteria by which the statutory standards are implemented; thus, the statute is not at risk of becoming out-of-date simply because the medical consensus about the methods to be used – such as those distilled in the "Guidelines" of the Commission's medical consultants – will certainly change. There is, however, nothing to indicate that those changes will not occur within the boundaries set by the statute, namely, the two standards of irreversible cessation of cardiopulmonary functions or irreversible cessation of brain functions.

B. *Conceptual Soundness*

Second, is the proposed UDDA analytically sound? Here, too, I believe the answer is affirmative. The relevant features are that it insists on uniformity among people and situations and that it measures death on the basis of the organism as a whole, not merely of particular parts.

The need for uniformity has been doubted by some writers, but the heart of their objections is, in effect, that no "definition" is needed at all [13]. In many contexts this notion is sensible; in this view, each situation in which it could be relevant to determine that a person has died should be governed by a particular statement of criteria for declaring death for that purpose. There are several objections to this approach regarding the present topic, however. First, no one has seriously proposed any need for varying "definitions" in varied contexts ([9], p. 645). There is no ground comparable to that which leads the law to declare a corporation is a "person" for purposes of owning property, while using a different definition of "person" (one that excludes corporations) for voting purposes.

More important, confusion rather than clarity would be introduced if different physical conditions were called by the same name, "death." When, for example, it is suggested that perhaps organs should be removed from patients (who have agreed in advance) at a point in the dying process when it would not otherwise be appropriate to bury and mourn the body, what is really being said is that organs should be removable before death has occurred ([10], pp. 107–108). Consequently, this special status – which could be called AFOR ("appropriate for organ removal") – should not be confused with "death". Keeping the categories distinct is not only theoretically sounder but would also facilitate using special procedures before acting in cases of AFOR – or other situations of treatment termination – while such procedures would be unnecessary when one is dealing with a dead body.

Finally, there is no denying that the task of "defining" death differs from those situations in which a single thing – such as a creature of the law, like a corporation – is defined in varying ways for varying purposes. The death of a human being is a more complex matter; it cannot be rested *solely* on an arbitrary legal standard. In part, any definition "is admittedly arbitrary in the sense of representing a choice," as the President's Commission stated in defending the view that the brain's **function** is more central to human life than are other necessary organs (such as the skin, liver, and so forth) ([26], p, 35). But the societally determined view of what constitutes death is not "arbitrary in the sense of lacking reasons" ([26], p. 35). Rather, societies have always had reasons for recognizing that death occurs, and these have typically been premised on noting that it is

a phenomenon which we think we observe in the life cycle of all beings. Of course, the definition can no more be "clean" of the cultural context in which it will operate than it is of the thing itself ([9], p. 643).

In other words, the "cultural context" of the standards for determining death includes the generally held view that human death, like the death of any animal, is a natural event. Even in establishing their "definition", members of our society act on the basis that death is an event whose existence rests on certain criteria recognized rather than solely invented by human beings. Indeed, there are good reasons for taking as the starting point the premise that what is being "defined" for humans ought not to diverge at its core from a "definition" that would be used for saying that any other animal is dead, while recognizing that the special qualities that are valued in humans *might* require distinctive features not germane to "lower" species. The UDDA, by supplying a definition that can be employed in all the myriad legal and social contexts in which it is important to know which human beings are alive and which are dead, meets the first part of the test of conceptual soundness.

The second conceptual question is whether there are distinctive human features that would lead to a radically different basis for determining human death from the death of other organisms. This is a question of what constitute necessary and sufficient conditions. On the one hand, circulatory and respiratory functions appear to be *necessary* functions for human life; though their origin may be mechanical (rather than from natural cardiac and pulmonary mechanisms), they must be present for any and all other organic functions (including those of the brain). On the other hand, does the mere existence of these functions amount to human life? Clearly not if their origin is totally artificial and not a result of any operations of the brain, otherwise a functioning heart-lung machine by itself (or a decapitated corpse hooked up to a ventilator) would be a living person. But is the persistence of spontaneous cardiopulmonary activity in a comatose patient *sufficient* to say that human life remains?

The Commission's affirmative answer to this question meets with near universal agreement among physicians and other health professionals, among legislators and other public officials, and among members of the public encompassing a very wide spectrum of cultural and religious viewpoints. Yet it continues to draw objections from some philosophers as conceptually unsound. Before explaining further how the Commission arrived at its conclusion, I want to emphasize two points that I think account for a good deal of the disagreement.

The first concerns the implications of the view that a spontaneously respiring human being is a living person. This conclusion does not imply that all people who meet this criterion are equivalent for all purposes; there are contexts in which it may make sense to draw distinctions based on the degree of mental functioning, so that a retarded or senile or comatose person is regarded as having different rights and responsibilities than a person capable of more "normal" mentation. Indeed, in its subsequent report on decisionmaking about life-sustaining treatment, the President's Commission concluded that a patient (like Karen Quinlan) who is reliably diagnosed to have permanently lost consciousness is not owed as vigorous medical care as are patients with better prognoses. Since their disability is "total" and no return to "even a minimal level of human functioning is possible," such patients' only interest in continued maintenance is "the very small probability that the prognosis of permanence is incorrect" ([25], p. 182).[2] Consequently, one ought not to assume that the decision to continue the traditional understanding of death (albeit using the new neurological methods when the old circulatory and respiratory ones are inapplicable) has any necessary implications regarding our attitudes toward these "diminished persons" and our obligations toward them. Concerns about "too much" or "inappropriate" treatment of seriously ill patients are important components of the current reexamination of the limits of medicine, but they carry no special weight in the "definition of death" debate.

A second reason for confusion about the necessary standards for determining death is that people are sometimes too literal in equating a particular organ's persistence or cessation with life or death. A reductionist approach of this sort can, of course, lead to silly results; but however telling it may seem rhetorically to say that "if one small reflex were all that were left of an organ's functioning that could not be human life," still, as a practical matter, this seemingly odd position would have to be tolerated if the only way to diagnose the existence of more significant functions involves clinical measures that also encompass the less significant ones (especially if their isolated persistence would be a very rare and temporary event).

The origins òf this confusion may be linguistic or they may be in the somewhat reductionist analysis that results from paying excessive attention to the functioning of various organ systems in formulating competing "definitions" – in other words, confusion between what is being measured and the means of measurement. The term "brain death" illustrates this: it is often used in a way that suggests the speaker means "the death of the brain," when what is actually at issue is "the death of a human being, as determined by an examination of brain functioning."

Both these confusions – about the implications of a definition and in the false equation of the thing being measured with the means of measurement – became apparent to the Commission during its study, just as they continue to intrude on discussions today. Their major manifestation is in the debate over the appropriateness of the two major views about brain functioning relative to death – the so-called "neocortical" (or "higher brain") and "whole-brain" definitions.

The former view can be defended on two alternative grounds. Some philosophers point to those activities, such as thinking, reasoning, feeling, and human intercourse, that make human beings distinctive. Without higher brain functions these attributes of "personhood" disappear; hence, death of the person occurs when the neocortex ceases functioning, even though a breathing body remains.

Other philosophers have argued that a patient's "personal identity" disappears when the higher brain functions are lost because one's identity depends on the persistence of mental processes. These processes supply the consistency in one's reactions to the world and one's memory of it; they are subject to gradual change but when a radical, destructive event occurs, the changes may be so grave as to destroy one's identity.

The Commission rejected these "higher brain" formulations for theoretical as well as practical reasons. Nothing approaching agreement exists about what things are essential to "personhood," or about how to solve the philosophical problem of identity. Further, the implication of the neocortical standards is that a person in a persistent vegetative state (like Karen Quinlan), who retains lower brain functions and hence breathes spontaneously, is as dead as any other corpse. The Commission – in line with our Judeo-Christian heritage and most Americans' current beliefs – rejected "this conclusion and the further implication that such patients could be buried or otherwise treated as dead persons" ([26], p. 40).

Moreover, the scientific and clinical knowledge necessary to implement a "higher brain" standard is lacking.

[I]t is not known which portions of the brain are responsible for cognition and consciousness; what little is known points to substantial interconnection among the brain stem, subcortical structures and the neocortex ([26], p. 40).

Even when aspects of consciousness can be linked with particular sites in the cerebral cortex, the methods for assessing irreversible cessation in the functioning of these sites are not yet sufficiently accurate to provide for certain diagnosis of death.

In contrast, the "whole-brain" formulation was found by the Commission to be conceptually sound, in addition to providing a well-verified basis for diagnosing death (as described above). This formulation recognizes that the heart, lungs, and brain have special significance in maintaining the integrated functioning of the human organism as a whole because, unlike other "vital" organs (liver, kidneys, skin, etc.), "their interrelationship is very close and the irreversible cessation of any one very quickly stops the other two" ([26], p. 33). Whether this interrelationship is regarded as basic to a living organism, or, alternatively, stress is placed on the centrality of the whole-brain to the organism's regulation and integration, is not really important in policy terms. On the former view, the irreversible loss of cardiopulmonary functions amounts to death because it shows directly that the integrated functioning of the three essential organ systems has ended. On the latter view, "the heart and lungs are not important as basic prerequisites to continued life but rather because the irreversible cessation of their functions shows that the brain has ceased functioning" ([26], p. 34). Either is consistent with the formulation of standards in the UDDA.

Like the "higher brain" view, the "whole-brain" standard is subject to criticism. The strongest argument, in my view, is that even though intensive medical support for patients without any brain functions cannot truly replace everything that a functioning brain stem achieves, and hence such patients cannot be maintained indefinitely, this proves only that such patients are dying, not that they are dead. The Commission found more persuasive (a) the argument that what remains is not a human being but "merely a group of artificially maintained subsystems" ([5], p. 391), and (b) the observation that, howsoever hard it is to discern precisely, a patient without any brain functions is missing – in a way that a person without neocortical functions is not – that cluster of attributes essential to an organism's responsiveness to its internal and external environment.

The startling contrast between bodies lacking *all* brain functions and patients with intact brain stems (despite severe neocortical damage) manifests this. The former lie with fixed pupils, motionless except for the chest movements produced by their respirators. The latter cannot only breathe, metabolize, maintain temperature and blood pressure, and so forth, *on their own* but also sigh, yawn, track light with their eyes, and react to pain or reflex stimulation ([26], p. 35).

In the end, the Commissioners candidly acknowledged that the theoretical arguments were not conclusive, one way or the other. Nonetheless, both views of the "whole-brain" formulation lead to the same statutory standard (as set forth in the UDDA), and anyone meeting that standard would also be dead according to the "personhood" and "personal identity" theories that undergird the neocortical view of death. It is here that my view that the Commission's report should be judged primarily on grounds of public policy, rather than on philosophical grounds, has the clearest consequence: it is acceptable to bring conceptual debate to a close in order to adopt a needed public policy, so long as one has reached a coherent and sensible result, albeit not one that resolves all philosophical doubts.

C. *Public Accessibililty*

The final two objectives for legislation, though crucial, can be dealt with relatively briefly. It is important that a statute that may affect every ordinary person and that controls a matter as central to people's relations with each other (as parent, spouse, and friend) and with the state (as taxpayer, voter, beneficiary, etc.) as the "definition" of death that it be comprehensible to the general public as well as the medical profession.

To enhance public understanding, the President's Commission concluded that a statute should have two features. First, it ought to speak in terms that laypeople can comprehend in light of their present experience and understanding. For most people, death is signified by an absence of breathing and heartbeat. Since that will continue to be true for the foreseeable future for the overwhelming majority of people (who are not going to die under circumstances in which artificial support precludes reliance on the traditional signs of death), the Commission concluded that the statute ought explicitly to acknowledge "irreversible cessation of circulatory and respiratory functions" as one basis for determining that death has occurred.

Second, to educate the public the statute ought to show the parallel between the cardiopulmonary standard and the brain-based standard. Linking these two closely, as the UDDA does, demonstrates that they are two windows through which the *same* phenomenon can be viewed – whereas separating them by elaborate phraseology, as the progenitor statute in Kansas did, creates the misimpression that there are two separate and distinct phenomena ([10], pp. 108-111).

Thus, by avoiding unnecessary, radical changes the UDDA is more publicly accessible. Not only does it continue to recognize the familiar heart-lung standard, but it does not move to the neocortical standard that would imply that the traditional definition had been in error all along (even before the advent of respirators, etc.). The public and its representatives have enough difficulty today deciding what treatment decisions are acceptable for seriously ill, dying patients without being told that such patients (like Karen Quinlan) are really "dead" albeit still breathing.

D. *Legal Simplicity*

The surest way to achieve the final objective of legal simplicity is for a statute to be focussed on the issue at hand and neither be overbroad nor attempt to encompass more than the problem area needing attention. In the case of the UDDA, the problem being addressed is how to make clear that a medical determination that death has occurred will be acceptable legally if it is based on measurements of brain functioning – or rather, their absence – when the traditional life-signs have been obscured by the use of artificial means of support. There is no need for the statute to command the medical, social, or religious behavior that should follow on determination that a person has died – and the Commission's proposal does not address such matters. The statute need not, in fact, "define" death – or life – at all, since that is not really the issue, merely the legal recognition of a necessary alternative means of diagnosing death.

Further, although a statute of the type drafted by the Law Reform Commission of Canada, which propounds the irreversible cessation of brain functions as the *sole* "definition," would appear to be simpler than one that recognizes cardiopulmonary cessation as an alternative, in fact it is not. Such a statute would require that conscientious physicians perform tests of brain functioning even in clear cases of cardiopulmonary cessation or, as in the Canadian proposal, the statute itself would have to recognize explicitly that the brain-based standard can also be met "by the prolonged absence of spontaneous circulatory and respiratory functions" ([20], p. 25). Nor is this concern about greater complexity, and even unclarity, merely hypothetical. In the United States, the point is illustrated by the ABA proposal, which set forth the single standard of "irreversible cessation of total brain function." Because this single standard might appear to be irrelevant to death-determination in most cases – in which the diagnosis will depend on measuring cardiopulmonary cessation, not brain cessation – several states amended the statute to permit determination made by "other usual and customary procedures," without explaining these procedures or their relationship to the brain-based standard.

IV. THE FUTURE

Since 1980, when the UDDA was recommended, 27 states have enacted statutes, some replacing prior legislation, most acting for the first time. The overwhelming trend is toward the UDDA, with 24 legislative adoptions and two explicit "adoptions" by the highest court of a state (which in both cases also urged the state legislature to act, which the Indiana legislature did five years later); 46 states now have an updated legal "definition" of death.[3]

Nonetheless, the pace of statutory adoptions has slowed and we are still quite far short of uniformity on this issue. The recommendations of the President's Commission had an initial impact, but beyond distributing its report the Commission did not believe it appropriate for it to become a force actively seeking action at the state level, except when invited by state legislative committees; moreover, the Commission had to turn its attention to its nine other reports. Furthermore, other than transplant groups, there is no natural constituency for the UDDA – after all, outside Chicago, dead people do not vote.

And, as the Commission concluded, it is a mistake to link the statute defining death to the transplant statute – as Connecticut and Illinois have done, with unfortunate results. There is no question that, as an historical matter, public and professional attention first focused on the issue of "updating" the standards for determining death because of organ (particularly heart) transplantation, which was perceived to be impeded by the existing common law. Yet from the testimony presented to the Commissioners and from the empirical study we conducted, it became apparent that even in settings where transplantation often takes place, only about 15% of those patients who are declared dead based on neurologic criteria become organ donors. In determining the death of a potential donor, the objective must, of course, remain to make a correct diagnosis and no step may be taken that compromises the donor's well-being to achieve benefit for third parties. Since the status "organ donor" is an arbitrary, rather than an inherent one, and is subject to manipulation by others, creating a separate standard of death for this group alone opens the door to abuse. Moreover, a special standard only for organ donors would fail to provide for 85% of respirator-supported cases that are beyond coma, leaving unchanged the present limbo of pointless treatment of these dead bodies.

The potential for confusion created by a separate standard was illustrated a couple of years ago in Connecticut, which adopted a "brain death" statute in 1979 but attached it to its anatomical gift act. A case arose of a young woman who had lost all brain functioning as a result of general anesthesia during dental surgery. She was not a potential organ donor, however, and the court held that she was therefore not dead – although she could have been so declared had she been a potential donor. Only when the public prosecutor reversed his original stand – for this one case – and agreed not to bring charges, did the physician feel he could follow through on his medical diagnosis of death and cease artificial support [14], [15].

Besides the lack of organized support, the failure of the remaining American jurisdictions to enact the UDDA derives not from any legislative preference for the radical step urged by proponents of "higher brain" or neocortical "definitions," but from opposition, primarily from a few religious groups, that misinterprets the intent and effect of the proposed statute. Probably the most groundless opposition is that based on the contention that loss of all brain *functions* is not enough for death, which should only be declared on proof of *organic destruction* [7]. The premise is that "until an organ has been destroyed there is always the *possibility* that it might resume functioning" ([26], p. 75). Of course, this standard has not been applied in diagnosing death on cardiopulmonary grounds. It is enough that a determination of *irreversible* loss of functions take account of current medical capabilities – indeed, the methods for both measuring and reversing loss of cardiac functions have increased dramatically over the past thirty years. "Since the evidence reviewed by the Commission indicates that brain criteria, properly applied, diagnose death as reliably as cardiopulmonary criteria, the Commission [saw] no reason not to use the same standards of cessation for both" ([26], p. 76).

A second objection to the UDDA – this one based on policy implications rather than doubts about medical science – is that any statute that would permit death to be declared on a respirator-supported body encourages euthanasia (first passive, and eventually active). This view, advanced by some right-to-life groups, has been most effectively rebutted by theologians and activists of the same ideological outlook. For example, a pro-life Catholic scholar argues that

a correct definition of death, if it would eliminate some false classifications of dead individuals [as being] among the living, could relieve some of the pressure for legalizing euthanasia – in this case, pressure arising from a right attitude toward individuals really dead and only considered alive due to conceptual confusion ([16], p. 61).

Similarly, Dennis J. Horan, President of American Citizens United for Life, declares that a statute incorporating irreversible cessation of total brain functions would be "beneficial" and would "not undermine any of the values we seek to support" ([18], p. 26). Catholic organizations, such as state Catholic Conferences, have supported legislative adoption of the UDDA in several states.

The final source of opposition arises from a disagreement in the orthodox Jewish community. Although the leading view is that complete cessation of brain function provides a basis for declaring death (by analogy to decapitation), some rabbis interpret the Jewish texts as requiring cessation of corporal blood flow, even though it is being artificially maintained in a manner not contemplated when those texts were set down ([26], p. 11).

There are two responses to the concern of this latter group about "premature" declarations of death. The first is that in some cases personal beliefs must yield to the needs of the collectivity – just as objections to autopsy must give way in appropriate cases to the state's investigatory requirements. The answer to the question whether a person is alive or dead is so central to many matters of public concern (as reflected throughout the civil and criminal law) that it cannot be left to individual preferences, no matter how sincerely held. Some who once urged a general "conscience clause," which would permit an individual to specify the standard to be used for determining his or her death ([29], pp. 72–76), seem now to recognize the chaos that could result were people free to declare that their "death" had occurred when senility arrived or, conversely, only upon total putrefaction ([8], pp. 356–357). Instead, the suggestion has been narrowed: people should be allowed to choose from among a few socially sanctioned alternatives. While eliminating the worst kinds of mischief, this proposal still proceeds on the false premise that society would find it acceptable to treat two physically identical individuals in radically different ways (i.e., one as dead and the other as alive).

More important, by restricting the range of "definitions," the proposal contradicts its own major selling point, namely, that "defining" death is a matter best left to individual conscience. There should, of course, be opportunity for choice regarding death – in the sense of the process of dying and the extent of artificial intervention in bodily processes. But, as has already been established regarding those who are concerned with what they regard as excessive medical maintenance of patients who have lost cognitive capacity, a statute on the standards for determining death issues no commands – one way or the other – about whether or when treatment may be ceased. Thus, if Orthodox Jews disagree with a determination that death has occurred in a body with artificially maintained respiration and circulation but no brain functions, they are not precluded from continuing to provide medical support (just as a potential kidney donor is maintained after death occurs).

In sum, the remaining impediments to the adoption of the UDDA arise from misinterpretations of the statute's premises and probable consequences. Like the philosophical arguments over the loss of "personhood" and "personal identity" that some believe occur when neocortical functions cease permanently, these objections ought not to delay enactment of this sensible and important law, which would permit our country's long-standing understanding of death, based on the Judeo-Christian tradition as well as on basic biological principles, to be restated in light of current medical capabilities.

University of Southern California
Los Angeles, California

NOTES

[1] Total cessation of heartbeat occurs within hours or a few days in most respirator-supported, "brain dead" adults, though it may persist in a few for 10–14 days given vigorous support [6]. One recent report of the survival of cardiopulmonary functions for 68 days of mechanical ventilation and extensive medical support does not contain sufficient data to establish whether the diagnosis of "brain death" met the criteria suggested by the medical consultants to the President's Commission [24].

[2] The Commission therefore concluded that the extent of care provided to such patients ought to be determined by the wishes of, and the direct and indirect consequences for, others, including the patients' relatives, caregivers, and society at large ([25], pp. 183–186).

[3] The following states have adopted laws on the determination of death that recognize cessation of total brain functions. B = ABA proposal (1975); C–K = Capron-Kass proposal (1972); K = Kansas Model (1970); M = AMA proposal (1979); N = nonstandard; UBDA = Uniform Brain Death Act (1978); and UDDA = Uniform Determination of Death Act (1980):

N	Alabama Code §§22–31–1 to 22–31–4 (1979) (permits use of other, unspecified medically accepted standards; when determination is based on cessation of brain function or when part of body is to be used in transplantation, must have second confirmation of death; limits liability for civil/criminal actions).
N	Alaska Stat. §09.65.120[2] (similar to UDDA; explicitly permits pronouncement of death before removal of artificial supports).
N	Arizona (see State v. Fierro, 124 Ariz. 182, 603 P. 2d 74 (1979); homicide law).
UDDA	Arkansas Stat. Ann. §§82–537 to 82–538 (1985).
UDDA	California Health & Safety Code §§7180–7183 (1982) (includes sections on confirmation and procedures for UAGA, and on patient records).
UDDA	Colorado Rev. Stat. §12–36–136 (1981).
N	Connecticut Gen. Stat. Ann. §19a–278(b) & (c) (1979) (part of UAGA).
UDDA	Delaware Code, tit. 24, §1760 (1986).
UDDA	District of Columbia Code Ann. §6–2401 (1982).
N	Florida Stat. §382.085 (1980) (determination must be made by two physicians – one a neurologist, neurosurgeon, internist, pediatrician, surgeon, or anesthesiologist; limits liability).
UDDA	Georgia Code Ann. §31–10–16 (1982).
N	Hawaii Rev. Stat. §327–C–1 (1978, am. 1984).
UDDA	Idaho Code §54–1819 (1981).
B	Illinois Ann. Stat. Ch. 110 1/2, §302 (1975) (part of UAGA).
UDDA	Indiana West's Ann. Ind. Code §1–1–4–3 (1986).
C–K	Iowa Code Ann. §702.8 (1976).
UDDA	Kansas Stat. Ann. §§77–204 to 77–206 (1984).
N	Kentucky Rev. Stat. §446.400 (1986).
C–K	Louisiana Rev. Stat. Ann. §9:111 (1976).

UDDA   Maine Rev. Stat. Ann. §§22–2811 to 2813 (1983).
UDDA   Maryland Ann. Code HG §§5–201 to 202 (1972.a.m.1983).
N   Massachusetts (see Commonwealth v. Golston, 373 Mass. 249, 252, 366 N.E.
        2d 744, 747, (1977), cert. denied, 434 U.S. 1039 (1978; homicide law).
C–K   Michigan Compiled Laws Ann. 333.1021 to 333.1024 (1979) (requires pro-
        nouncement of death before terminating artificial supports; applies to trials of
        all civil and criminal cases).
UDDA   Mississippi Code 1972, §§41–36.1 to 41–36.3 (1981).
N   Missouri Vernon's Ann. Mo. stat. §194.005 (1982).
UDDA   Montana Rev. Codes Ann. §50–22–101 (1983).
N   Nebraska (see State v. Meints, 212 Neb. 410, 414, 322 NW 2d 809, 812 (1982);
        homicide law).
UDDA   Nevada Rev. Stat. §451.007 (1979).
UDDA   New Hampshire R.S.A. §§141–D:1 to 141–D:2 (1986).
K   New Mexico Stat. Ann. §12–2–4 (1978).
(C–K)   New York (see People v. Eulo, 63 N.Y.2d 341, 346, 482 N.Y.5.2d 436, 438
        (1984)).
N   North Carolina Gen. Stat. §90–323 (1979).
M   Ohio Rev. Code Ann. §2108.30 (1982) (mentions brain stem and artificial
        support).
UDDA   Oklahoma Stat. Ann. tit. 63, §§3121 to 3123 (1986).
UDDA   Oregon Rev. Stat. §432.300 (1987).
UDDA   Pennsylvania Stat. Ann. 35 §§10201 to 10203 (1982).
UDDA   Rhode Island General Laws §23–4–16 (1982).
UDDA   South Carolina Code 1976, §44–43–450 and §44–43–460 (1984).
UDDA   Tennessee Code Ann. §68–3–501 (1982).
C–K   Texas Rev. Civ. Stat. Ann. art. 4447t (1979).
UDDA   Vermont Stat. Ann. tit. 18, §5218 (1981).
K   Virginia Code §54–325.7 (1973, am. 1979).
(UDDA) Washington (see In re Bowman, 94 Wash. 2d 407, 421, 617, P.2d 731, 738
        (1982)).
UBDA   West Virginia Code §§16–10–1 to 16–10–3 (1980).
UDDA   Wisconsin Stat. §146.71 (1982).
UDDA   Wyoming Stat. §§35–19–101 to 35–19–103 (1985).

BIBLIOGRAPHY

1. Ad Hoc Committee of the Harvard Medical School to Examine the Definition of
   Brain Death: 1968, 'A Definition of Irreversible Coma', *Journal of the American
   Medical Association* **205**, 337–340.
2. Alexander, M.: 1980, 'The Rigid Embrace of the Narrow House: Premature Burial
   and the Signs of Death', *Hastings Center Report* **10**, 25–31.
3. Arnold, J. *et al.*:1968, 'Public Attitudes and the Diagnosis of Death', *Journal of the
   American Medical Association* **206**, 1949–1954.

4. Barclay, W.: 1981, 'Guidelines for the Determination of Death', *Journal of the American Medical Association* **246**, 2194.
5. Bernat, J. *et al.*:1981, 'On the Definition and Criterion of Death', *Annals of Internal Medicine* **94**, 389–394.
6. Black, P.: 1978, 'Brain Death', *New England Journal of Medicine* **299**, 338–344.
7. Byrne, P. *et al.*: 1979, 'Brain Death: An Opposing Viewpoint', *Journal of the American Medical Association* **242**, 1985–1990.
8. Capron, A.: 1978, 'Legal Definition of Death', *Annals of the New York Academy of Science* **315**, 349–359.
9. Capron, A.: 1973, 'The Purpose of Death: A Reply to Professor Dworkin', *Indiana Law Journal* **48**, 640–646.
10. Capron, A. and Kass, L.: 1972, 'A Statutory Definition of the Standards for Determining Human Death: An Appraisal and a Proposal', *University of Pennsylvania Law Review* **121**, 87–118.
11. Conference of Royal Colleges and Faculties of the United Kingdom: 1979, 'Memorandum on the Diagnosis of Death', in Working Party of the United Kingdom Health Departments, *The Removal of Cadaveric Organs for Transplantation*: *A Code of Practice*, London, pp. 32–36.
12. Cranford, R. and Smith, H.: 1979, 'Some Critical Distinctions Between Brain Death and Persistent Vegetative State', *Ethics in Science and Medicine* **6**, 199–209.
13. Dworkin, R.: 1973, 'Death in Context', *Indiana Law Journal* **48**, 623–639.
14. Fabro, F.: 1981, 'Bacchiochi vs. Johnson Memorial Hospital', *Connecticut Medicine* **45**, 267–269.
15. Fabro, F.: 1981, 'The Bacchiochi Case – Continued', *Connecticut Medicine* **45**, 334–335.
16. Grisez, G. and Boyle, J.: 1979, *Life and Death with Liberty and Justice: A Contribution to the Euthanasia Debate*, University of Notre Dame Press, Notre Dame, Ind.
17. Heiden, J. *et al.*: 1979, 'Severe Head Injury and Outcome: A Prospective Study', in Popp, A. *et al.* (eds.), *Neural Trauma*, Raven Press, New York.
18. Horan, D.: 1980, 'Definition of Death: An Emerging Consensus', *Trial* **16**, 22–28.
19. Kübler-Ross, E.: 1969, *On Death and Dying*, Macmillan, New York.
20. Law Reform Commission of Canada: 1981, *Report on the Criteria for the Determination of Death*, Minister of Supply and Services Canada, Ottawa.
21. Levy, D. *et al.*: 1981, 'Prognosis in Nontraumatic Coma', *Annals of Internal Medicine* **94**, 293–298.
22. McWhirter, N. (ed.): 1981, *The Guiness Book of World Records*, Bantam Books, New York.
23. Medical Consultants on the Diagnosis of Death to the President's Commission for the Study of Ethical Problems in Medicine and Biomedical and Behavioral Research: 1981, 'Guidelines for the Determination of Death', *Journal of the American Medical Association* **246**, 2184–2186.
24. Parisi, J. *et al.*: 1982, 'Brain Death with Prolonged Somatic Survival', *New England Journal of Medicine* **306**, 14–16.
25. President's Commission for the Study of Ethical Problems in Medicine and Biomedical and Behavioral Research: 1983, *Deciding to Forego Life-Sustaining Treatment*, Government Printing Office, Washington, D.C.

26. President's Commission for the Study of Ethical Problems in Medicine and Biomedi-
    cal and Behavioral Research: 1981, *Defining Death*, Government Printing Office,
    Washington, D.C.
27. Public Law 95–622 (1978), codified at 42 U.S.C. §300v–1(a)(1)(B) (Supp. 1982).
28. United Trust Co. v. Pyke, 427 P.2d 67 (Kan. 1967).
29. Veatch, R.: 1977, *Death, Dying and the Biological Revolution: Our Last Quest for
    Responsibility*, Yale University Press, New Haven.

ROBERT M. VEATCH

WHOLE-BRAIN, NEOCORTICAL, AND HIGHER BRAIN RELATED CONCEPTS

The organizers of this conference have chosen to contrast the *whole-brain* and *neocortical* definitions of death. It is not clear how precisely they are using the terms. Presumably, to be precise, a whole-brain oriented concept of death is one that measures death by the loss of all brain structure or functions, while a neocortical concept of death is one that measures death by the loss of all neocortical structure or functions.

The original critique of the whole-brain oriented definition of death was rooted in the point that it seems bizarre to consider a person alive simply because some utterly trivial brain stem reflex arc happens to be intact. Surely, whatever we mean by death, so the argument goes, some brain function might be retained while still considering the human being with only that function dead. This led some to conclude that a person could really be dead if there were irreversible loss of just neocortical brain function, thus permitting one diagnosing death to ignore the extraneous brain stem reflex. Several problems arise with the neocortical position, however.

One of the first challenges came from those who realized that brain function at the cellular level might outlast organ system level function by a long period. The National Institute of Neurological Diseases and Stroke seemed to realize this when it decided to tolerate up to 0.2 microvolts of electron potential ([15], p. 983) while considering someone dead. Since EEG readings record neocortical activity, if there were any electrical activity, presumably something was alive. The obvious conclusion is that cellular level function, even at the neocortical level, does not count as anything significant any more than isolated lower brain activity did. Only supercellular or system-level functions count.

Upon reflection, however, even this solution does not prove satisfactory for those of us who are convinced that the critical functions are ones that are normally associated with the neocortex. It seems possible that even some system level neocortical functions might be considered trivial. If, for example, the critical functions were considered to be sensory, how would we assess someone who, due to a hypothetical freak accident, had an isolated section of motor cortex alive and functioning?

171

Richard M. Zaner (ed.), Death: Beyond Whole-Brain Criteria, pp. 171–186.
© 1988 *by Kluwer Academic Publishers.*

If stimulated, that cortex would control some movement of a hand or foot, but no consciousness or conscious control would be possible. Anyone who is sympathetic with the position that moves the search for a definition of death beyond the whole-brain and in the direction of the neocortex would have difficulty concluding that such a person were alive simply because he retained some system level neocortical motor activity. The neocortical conception of death, just like the whole-brain oriented one, may be too conservative.

On the other hand, it is possible that a neocortical conception of death may be too liberal. It is at least hypothetically possible that some functions normally associated with the neocortex (what we often refer to as "higher" functions) could be replaced by other brain tissues or – at some point in the future – by computers. Surely a person walking around with a microcomputer that permitted him to think and feel, but who had no living neocortical tissue, would still be alive. (This formulation is based on a related argument in Tomlinson [10].) What we are really interested in is not the presence of neocortical activity, but the presence of certain functions that, for want of a better term, we often refer to as higher brain functions. Exactly what these functions are is open to much further debate, but they would seem to include capacity to be conscious, to think, feel, and be aware of other people. It is striking to note that these are almost precisely the functions Henry Beecher, the Chairman of the Harvard Ad Hoc Committee, referred to when identifying why he thought a person with a dead brain should be considered dead ([1], pp. 2–4).

It seems clear that when people speak of a so-called neocortical concept of death, they really have in mind a concept of death in which a person is considered dead when he or she irreversibly loses certain supercellular functions traditionally associated with the neocortical portion of the brain. Should we someday be capable of preserving those functions in the absence of the neocortex, a person retaining those functions would surely be considered alive. Thus when we speak of neocortical death it is really a short-hand for the more cumbersome phrase: death based on the irreversible loss of supercellular functions traditionally related to the neocortex. To convey that nuance, I have for some years referred to a "neocortically oriented" or "neocortically related" concept of death rather than the simpler "neocortical death." By the same token I have tried to speak of a "whole-brain oriented" concept of death rather than the simpler "whole-brain death."

If, however, the term "neocortically oriented concept of death" conveys that death occurs when there is irreversible loss of all supercellular functions, but it is really plausible to hold that a person could be dead even while retaining certain supercellular neocortical functions (such as motor activity), then we need still another term. I have taken to speaking of "higher brain functions" to refer to the capacity to think, feel, be conscious and aware of other people, conceding that the term lacks precision. On this basis I shall contrast whole-brain oriented and higher-brain oriented concepts of death, leaving out of further consideration the somewhat different neocortical concept of death. I shall defend in this paper a higher brain concept. By that I mean that a person should be considered dead when there is an irreversible loss of higher brain functions, i.e., certain functions normally associated with the neocortex that include the capacity to be conscious, to think, feel, and be aware of other people.

The President's Commission consistently refers to a higher-brain concept of death rather than a neocortical one. It is the Commission's rejection of that concept that I address.

I. A CRITIQUE OF THE PRESIDENT'S COMMISSION

In its report on *Defining Death*, the President's Commission for the Study of Ethical Problems in Medicine and Biomedical and Behavioral Research ([9], p. 1) concluded that "death is a unitary phenomenon which can be accurately demonstrated either on the traditional grounds of irreversible cessation of heart and lung functions or on the basis of irreversible loss of all functions of the entire brain." With that conclusion very few people are in disagreement. That conclusion, however, finesses the really interesting philosophical and ethical question: can this unitary phenomenon of death also be demonstrated by the irreversible loss of some more limited set of so-called "higher brain" functions? It is clearly the Commission's intention to reject this possibility, although when it comes right down to it, it never offers a single argument that actually supports that conclusion. It does present three pages summarizing the higher brain position. It concludes that section, however, with a nonsequitur: "Thus, all the arguments reviewed thus far are in agreement that irreversible cessation of *all* brain functioning is sufficient to determine death of the organism" ([9], p. 41). The real question today, however, is not whether this is a sufficient condition, but whether it is a necessary condition. The advocates of higher brain formulations have argued that the irreversible loss of higher brain function is also sufficient to determine the death of the organism.

As early as the early 1970s a number of people were aware that a great deal of what some people took to be essential to human existence could be lost without the destruction of the entire brain or even the loss of all brain functions. The critical scientific paper, at least for those of us involved in the Hastings Center's Death and Dying Research Group (the group which stimulated much of the philosophical and public policy reflection on this subject) was that by Brierley and his colleagues [2] in 1971. By 1973, in a paper presented at a Foundation of Thanatology symposium, I had reached the conclusion that the whole-brain oriented conception of death was not sufficiently precise (published in 1975 [13]). Others seemed to be coming to a similar conclusion ([7], p. 133; [5], pp. 587–590). This was apparently sufficient for the President's Commission to treat the higher-brain related formulation of a concept of death with some seriousness. They offered what I count as four points against a higher-brain formulation and in favor of one more oriented toward the irreversible loss of all brain functions. As we shall see, not one of them really counts as an argument.

1. *The Lack of General Agreement*

The Commission begins its discussion by reducing the higher-brain formulation to one involving personhood such that loss of personhood is taken to be equated with death. The Commission then concludes that "crucial to the personhood argument is acceptance of one particular concept of those things that are essential to being a person, while there is no general agreement on this very fundamental point among philosophers, much less physicians or the general public" ([9], p. 39).

There are two problems with the Commission's discussion. First, it reduces the "higher-brain" formulation to one depending on a concept of personhood. Quite frankly, the philosophical discussion of the concept of personhood over the past decade has been less than illuminating. More to the point, there is no reason to assume that clarification of the concept of personhood must be achieved in order to consider someone dead based on loss of higher brain functions. I have never in my own discussions of the concept of death made my formulation dependent on the concept of the person. I have done so in part because of the confusion over that concept and in part because it has seemed quite possible to me that someone could have irreversibly lost personhood, whatever that may mean, and still be alive. If, for example, personhood is defined as the possession of the concept of "a self as a continuing subject of experiences and other mental states" [11], or as the possession of consciousness, reasoning ability, and self-motivated activity [16], making loss of personhood synonymous with death would make many people (such as the non-communicative senile) dead simply because they were no longer persons, while a consensus seems to exist that they should be treated as alive. I have preferred simply to avoid this thicket by attacking the definition-of-death problem without any reference to a theory of personhood. Let it be clear that I am not here attempting to avoid the difficult question of what characteristics are essential for treating someone as alive. In fact, it is precisely that question that must be answered and will be addressed below. Rather, I am claiming that that question can be dealt with without reference to the personhood debate or deciding what characteristics are essential to personhood, whatever that may mean.

The Commission seems to take the position that if there is no general agreement among philosophers (or physicians or the general public) on some concept, then it cannot be correct. The absence of consensus is surely not a sound argument against a position. In fact, there is no more of a consensus in favor of a whole-brain oriented formulation than there is in favor of a higher-brain oriented one. The only survey with which I am aware that would permit us to address this question seems to point to the conclusion that there is no majority support for *any* definition of death: whole-brain, higher-brain, or non-brain [4]. The Commission may have been meaning to argue that since there is no consensus the society should take the safer course and opt for the whole-brain oriented concept of death so that no one may be treated as dead who could by some philosophical views be alive. If that is the Commission's position, however, it is an argument in favor of adopting a heart and lung oriented concept of death, not a whole-brain oriented one.

The absence of a consensus may be a reason to be concerned about how to formulate a public policy, but not necessarily a reason for adopting the centrist alternative. I take the lack of consensus to be a good reason for adopting a more pluralistic and tolerant attitude about defining death, one that permits individuals to choose their own definition of death from among some small list of reasonable alternatives ([14], pp. 72–76; [12], pp. 316–317). However, it is wrong to conclude, as the Commission does, from the premise "There exists no consensus on a definition of death," that therefore "The whole-brain oriented definition should be chosen."

2. *The "Would You Bury a Breathing Body" Argument*

The Commission's second point against the higher-brain oriented formulations is equally a non-argument. The Commission points out that:

the implication of the personhood and personal identity arguments is that Karen Quinlan, who retains brain stem function and breathes spontaneously, is just as dead as a corpse in the traditional sense. The Commission rejects this conclusion and the further implication that such patients could be buried or otherwise treated as dead persons ([9], p. 40).

There are several problems here. It is true that by some concepts of person Karen Quinlan may not have been a person. I say "may not have been" because I, like the Commission, have doubts about what kind of empirical evidence it would take to demonstrate irreversible loss of personhood according to various concepts of personhood. Regardless, however, it would have been at least a mistake to have considered Karen Quinlan dead according to a neocortical concept of death. The neurological evidence was clear that she retained some neocortical function ([8], p. 654). If the neocortical conception of death is one that equates death to the irreversible loss of all neocortical functions, then Karen Quinlan could not have been dead by such a conceptualization any more than she could have been by whole-brain oriented conceptions. It is somewhat more difficult to judge whether she would have been dead according to a "higher-brain" conception of death. That would have depended on exactly what counted as the higher functions.

All of this may miss the point, however. The argument from our intuitions about whether we would bury someone who had lost all higher brain functions but retained spontaneous respiration is fallacious. It assumes that it is acceptable to bury people simply because they are dead – that is, corpses. Yet by any definition of death it seems clear that some people would not be buried immediately when they become dead. I would not want to bury a person who had been pronounced dead based on a whole-brain oriented concept of death if that person was still respiring on a respirator and his heart was still beating. On aesthetic grounds I would want to disconnect the respirator and let his heart stop before burial. I would also remove IVs, NG tubes, etc. The fact that we would not bury certain dead people until certain residual functions ceased is not a plausible argument against considering them dead. The fact that we would await cessation of spontaneous respiration for someone pronounced dead based on higher brain function loss is not a sound argument against the use of higher-brain related criteria for death.

3. *The "No Techniques are Available" Argument*

The third point made by the Commission against the higher-brain oriented formulation is that "at present, neither basic neurophysiology nor medical technique suffices to translate the 'higher brain' formulation into policy" ([9], p. 40). This is an important point, one that cannot be overemphasized. It may lead to the policy conclusion that in order to pronounce people dead (based on higher-brain conceptualizations of death) we must revert to the old whole-brain oriented criteria. The logic of such a move is that persons will be considered dead when they lose higher brain function, but that the only way we can know for sure that higher brain function has been lost is to demonstrate that all brain function has been lost. For the past eight years I have suggested precisely this conservative policy course ([14], p. 72). It is important not to overstate the difficulty, however. If we have real doubts that Karen Quinlan had irreversibly lost higher brain functions, that would have seemed to have had very critical implications for the decision that it is appropriate to withdraw respiratory and other medical support. If she was not dead because we could not determine that she had lost higher brain function, by the same token she may not have been a good candidate for stopping treatment. On the other hand, if we are sufficiently convinced of irreversible loss of mental capacity (the court actually referred to loss of "cognitive, sapient function"), maybe we can be sufficiently convinced to pronounce death based on loss of those functions.

Regardless of whether we want to take the safe course and not operationalize measures to diagnose irreversible loss of higher brain function, this decision is surely not a sound argument against the position that people ought to be considered dead when it *can* be determined that they have irreversibly lost higher brain function. It may be an argument leading to the conclusion that it is not urgent for policy purposes to distinguish between higher-brain and whole-brain formulations, but it does not count against a philosophical position that says humans are to be treated as dead when it has been established that they have lost higher brain functions.

4. *The "No New Concept" Argument*

Finally, the Commission argues that adopting a higher-brain oriented concept of death would be adopting a "new concept" of death and that "one would desire much greater consensus than now exists before taking the major step of radically revising the concept of death" ([9], p. 41). On its face, this is a strange point. It argues that a reason not to adopt a particular concept is that it would be new. Surely, one ought to adopt positions that are right and reject ones that are wrong regardless of whether they are new or old. Regardless of whether being new is a telling argument against a concept, the Commission's point requires a hidden premise that the whole-brain oriented concept is not new.

In order to understand what is happening here, some history is necessary. Since the early 1970s Alexander Capron, the Executive Director of the Commission, and I have disagreed on how to conceptualize the whole-brain oriented concept of death. For Capron, the whole-brain oriented conception requires no new understanding of death, no new conceptualization. He seems to believe that throughout history people have always held that death is equated with the irreversible loss of all brain function. They have simply measured this loss by looking at heart and lung activity.

However, the historical and linguistic evidence is to the contrary. My reading of the evidence is that people have traditionally meant by death the irreversible stopping of the flowing of bodily fluids – the blood and the breath. Only on this basis can people have said meaningfully that persons with dead brains were still alive. Only on this basis would it be necessary to have statutory and common law revisions of the concept of death. Only on this basis would we have to overturn the idea that people die when their heart and lung function stops. Shifting to the idea that death means the loss of brain function even though the fluids continue to flow required a major shift in conceptualization. It is simply wishful thinking to claim that the states that after years of often bitter struggle and debate have adopted brain oriented definitions of death have not made an important conceptual shift in their understanding of death.

If I am correct, then adopting either a whole-brain oriented concept or a higher-brain oriented concept is a major step, a radical revision of the concept of death. Capron's device of claiming that the whole-brain oriented concept has always been with us is implausible, even if for some reason it counts against a position that it is new.

This leads me to the conclusion that the Commission has in no way undercut the use of a higher-brain oriented concept of death. It may show that there has not been any general agreement in favor of a higher-brain concept (or any other concept for that matter), that we would not bury all dead persons immediately, that we may have to be overly conservative in measuring higher-brain related death, and that a higher-brain concept (like a whole-brain related concept) is a new concept. But none of these counts against (or for) the use of a higher-brain oriented concept. In order to defend the use of a higher-brain oriented concept, some additional work is necessary.

II. THE PHILOSOPHICAL UNDERPINNINGS OF THE HIGHER-BRAIN CONCEPT

One of the most sophisticated presentations that seems to support a higher-brain oriented formulation is that offered by Michael Green and Daniel Wikler in their article, "Brain Death and Personal Identity" [6]. They argue that death should be equated with the loss of personal identity. To the extent that personal identity can be equated with what we are referring to as higher brain functions or what is sometimes for convenience referred to as neocortical functions, their argument is an apparent defense of higher-brain or neocortical death.

While, as Green and Wikler suggest, the personal identity theory of the definition of death may have much to offer and while their conception of personal identity seems to lead to a conclusion that is similar to mine, I find there are problems with it. Their position, in contrast to my own, does seem, contrary to their claim, to require a theory of personhood (or at least personal identity). They are therefore subject to all the criticisms that suggest that a human being who has lost personhood or personal identity may nevertheless still be alive.

Consider the case of a man (following the Green and Wikler convention, let us call him Mr. Jones) who through a severe head trauma had clearly and irreversibly lost all personal identity according to the criterion suggested by Green and Wikler, but after months of medical administration regained consciousness and mobility (of course, not recalling anything of his past). Such events are at least conceivable; perhaps they have actually happened. Under the Green and Wikler personal identity conception of death, we would have to say that Mr. Jones was dead; that, therefore, assuming the normal behavioral correlates of calling him dead, Mr. Jones's beneficiaries would inherit his property; that his home and assets would no longer be his; and that, if he happened to be in public office, he would *automatically* be removed. We would face some very awkward moments. What, for instance, would we call the person who gets up and walks away from the hospital bed possessing a new personal identity? Who, in fact, would name him? How would he be supported – without any assets or even disability insurance or social security?

It seems clear to me that we should concede to Green and Wikler that Mr. Jones may have lost personal identity, but surely not that a death has occurred. Even if there had been a total loss of personal identity and a new personal identity created in the same body, I am convinced that we would have no trouble concluding that no death had occurred. Death is not merely the irreversible loss of personal identity.

I have consistently maintained that death should be the name we give to the condition under which it is considered appropriate to initiate a series of behaviors that are normally initiated when we call someone dead. I have referred to these as "death behaviors." We begin mourning in ways that were not included in anticipatory grief; we may decide that certain medical interventions should only be stopped at the time of death, in which case we cease those interventions; we begin the process of reading the will; life insurance companies pay off; social security and annuity checks cease; we begin referring to the person in the past tense ("Mr. Jones was a good man" rather than "Mr. Jones is a good man"); and, if the person were President of the United States, the Vice President is automatically elevated to the presidency. On the other hand, if we consider the person still alive, none of these behaviors is appropriate. We may engage in other behaviors – behaviors appropriate in response to a dying man or a seriously ill man, but we do not treat him as dead. We may, for example, anticipate the death; we may inquire about the location of a will, but not read it. If the person we consider still alive is President, we may initiate emergency procedures to have him removed from office, but we do not automatically have a new President.

Since the 1960s I have maintained that "death" is simply the name we give to the condition when these behaviors are considered appropriate. It may be that we now have to come to the conclusion that not all of them should occur at the same instant. In that case death as we know it would cease to exist. It would be replaced by a series of discrete events signalling the appropriateness of the various social behaviors. It seems, however, that society still is comfortable acknowledging a moment when all of these behaviors ought to occur. Whatever that moment is, it is not necessarily the same moment as the time when personhood, however defined, is lost, even irreversibly lost. It is not the moment when personal identity is lost, either. It surely is not the moment when all brain tissue is destroyed, as Byrne and his colleagues [3] would have us believe. That would commit us to having to measure the anatomical disintegration of all brain tissue. There is nothing illogical or unscientific about such a position, but it is both implausible and inconsistent with Judeo-Christian, Greek, and modern secular thought. We would not even insist on the irreversible destruction of all brain function. Whatever it is that leads us to the ethical conclusion that humans should not be treated as dead, it is not the mere presence of isolated cellular activity of the brain.

The critical factors are surely functions, and they are functions at the supercellular level. What functions these are is fundamentally a non-scientific question. It must be answered by appeals to religious or metaphysical world views about when we should stop treating human bodies the way we treat living beings. Increasingly we are convinced that these functions are not respiratory and circulatory. I cannot think of any possible reason why we would assign such significance to brain stem reflex arcs. We have since 1970 excluded spinal cord reflex arcs, which are not that different. Nevertheless, the position adopted by the President's Commission would force you to treat me as alive if my body had through some freak event preserved in my brain only the simplest gag reflex.

I do not want to be confused with my gagging. I would not want society to continue to treat me as if I were alive simply because my body could gag. By the same token, I would not want society to continue to treat me as if I were alive simply because my body retained some isolated perfused motor cortex capable of jerking my arm if stimulated properly. All of this leads me to the conclusion that I would want me considered dead in these circumstances. I even take it as a kind of insult that I could be confused with any of these trivial bodily capacities.

You may, at this point, press me for the underpinnings of my position. Can I give any reasons why I would want to be considered dead

rather than alive in these situations? My answer is that I can, but they
are reasons that are not likely to be persuasive to everyone in our
pluralistic and secular society.

I, like a great many in our society, stand in the Judeo-Christian
tradition. As such I maintain two things. First, I maintain that the
human is fundamentally a social animal, a member of a human community capable of interacting with other humans. Second, I maintain
that I am in essence the conjoining of soul and body – or to use the more
modern language, mind and body. If either one is irreversibly destroyed
so that the two are irretrievably disjoined, then I – this integrated entity
– no longer exist. What is critical is the embodied capacity for consciousness or social interaction. When this embodied capacity is gone, I am
gone. When there is no longer any capacity for consciousness, to think
and feel within a human body, then I am gone. Were capacity to interact
socially separable from the capacity for consciousness, I would have a
problem of deciding which is critical. But after many years of worry over
that question I am forced to the conclusion that they are not separable.
When this embodied capacity for consciousness or social interaction is
gone irreversibly, then, and only then, do I want society to treat me as
dead.

Now, however, is this simply an affirmation of a particular religious/
philosophical view with which others may disagree? I am sure that it is.
And the implications of that are significant. First, while my affirmation
that death behaviors are appropriate when, and only when, there is an
irreversible loss of embodied capacity for consciousness and social
interaction reflects a particular religious/philosophical view, so does any
other conceivable answer to the question of when death behavior is
appropriate. The answer that death behaviors are appropriate when and
only when respiratory and circulatory function ceases is surely a reflection of a strange animalistic view. Institutionalizing the circulatory and
respiratory oriented concept of death would be institutionalizing a
religious perspective just as surely as adopting a higher brain functions
position. Moreover, institutionalizing the view that bodily integration
maintained by the presence of *any* supercellular brain functions including lower brain functions is also adopting a particular religious/
philosophical position. It is the position that what really counts in
justifying treatment of a human as alive or dead is whether there is any
capacity to integrate bodily functions – respiratory function, brain stem
reflexes, and the like. That, like the heart and lung oriented conception

of death, is a reflection of animalism, the view that humans are nothing more than their animal functions.

Where does that leave us? If I am correct, the President's Commission has opted for a particular theological/philosophical view – that of the animalists who give highest priority to the capacity to integrate bodily functions and who are not all that different from those who would insist that heart and lung function is what counts. They have adopted a theological/philosophical position that is at odds with the Judeo-Christian tradition, but a theological/philosophical position nonetheless.

I am convinced that adopting the position more consistent with the Judeo-Christian tradition that affirms unity of mind and body is more plausible. I think that is sufficient reason for individuals to adopt it. It would, however, mean imposing a particular view on others – imposing it, for example, on some Jews and American Indians who apparently hold firmly to the heart-lung oriented concept of death, and to some animalists who hold firmly to the whole-brain oriented concept of death. As a matter of public policy (as opposed to personal conviction) I am unwilling to do that. I have therefore advocated a position that is more tolerant. Before the President's Commission I argued that this problem should be treated just like any other question of religious/philosophical pluralism. People should be permitted to examine their own religious and philosophical traditions and adopt the positions that are most plausible to them.

Picking a definition of death, of course, has behavioral consequences. In fact, I have argued, that is the only reason the question is of significant policy concern. In matters affecting behavior, we permit only limited religious and philosophical liberty. As a society that loves liberty we will generally tolerate conscientious behaviors as well as conscientious belief whenever we can, but we must set limits.

This, however, may be much less of a problem than we imagine. We must realize that many of the practical behavioral issues can be resolved independently of the definition of death. If, for example, someone holds that death occurs when and only when the heart stops beating, he still has available to him the option of refusing medical treatment so that the heart will stop. In an extreme situation, society could even insist that such treatment be stopped against the individual's will so that the heart stops. Death could be pronounced for that individual based on his personally held position that death occurs when the heart stops with virtually no social or economic implications. On the other hand, for the person who, like myself, held that death occurs at the time there is irreversible cessation of higher brain functions, death might be pronounced earlier than others in society would like. That, in itself, has virtually no policy significance, however. The alternative would be to require that such a person be considered alive until his whole-brain stops. In such a case it is widely recognized that the individual (now considered alive) or his agents would have the right to refuse further treatment so that he would be dead very soon in any case.

The only real problems created by the recognition that the choice of a definition of death should be left to individual conscience would be for that very small group at the extreme who might decide that they wanted themselves or the ones for whom they were agents considered dead when heart, lung, lower brain, and higher brain all continued to function, or for the other small group at the other extreme who might decide that they wanted themselves or the ones for whom they were agents considered alive even though heart, lung, lower brain, and higher brain all had ceased to function. Here, especially when decisions are being made for other parties, the behavioral implications are so great and the deviation is so serious, that society would have to place limits.

There is one objection to the pluralistic solution that needs to be addressed. It is sometimes argued that permitting individuals to choose, even within a limited range, would create confusion and administrative nightmares. The unconscious emergency room patient would be alive or dead depending on his views, and he would be in no condition to be asked what his views are. There is a simple solution to this problem, however. As a matter of public policy we could adopt any one of the three plausible concepts of death referred to by the Commission with the proviso that individuals could opt for one of the others. This is the way we handle related matters such as Uniform Anatomical Gift Act donations. We establish a default position and let those who are concerned opt for an alternative. In the case of the concept of death it might be acceptable to opt for the whole-brain oriented position as a middle of the road option. I would prefer we opt for the higher-brain oriented position for the default. The choice would not matter much if those who objected had the opportunity to opt out.

My conclusion is that the President's Commission has made two mistakes. First, it has mistakenly rejected the higher-brain oriented concept of death as the most plausible. For a country that stands so close to the Judeo-Christian tradition to reject the position favored by that tradition and by most contemporary scholars within that tradition seems odd. Second, at the policy level it has missed the obvious and traditional way of resolving policy conflicts when matters of religious and philosophical variation are at stake. Permitting individuals to exercise their consciences based on their religious and philosophical beliefs would make the most sense. It would create virtually no adverse impacts on others within the society. Only in extreme cases would limits have to be placed. Those limits could easily be set. The Commission's recommendation, as it stands, manages to violate the convictions of both a conservative group of Jews and others who could be considered dead when they would want to be considered alive, and a more liberal group of Jews, Christians, and secular thinkers who could be considered alive when they would want to be considered dead. Were there a good reason to force a position onto these people against their will, perhaps it could be tolerated, but there is not. Had the Commission adopted one of the positions – preferably the most plausible higher-brain oriented concept – and then permitted individuals to dissent, ethical and religious freedom would have been preserved, mental anguish would have been prevented, and greater philosophical clarity would have been obtained, all at no significant social and economic costs.

The Kennedy Institute of Ethics
Georgetown, University
Washington, D.C.

BIBLIOGRAPHY

1. Beecher, H. K.: December 1970, 'The New Definition of Death, Some Opposing Views', Paper presented at the meeting of the American Association for the Advancement of Science.
2. Brierley, J. B., Adam, J. A. H., Graham, D. I., and Simpson, J. A.: 1971, 'Neocortical Death after Cardiac Arrest', *Lancet* **2**, 560–565.
3. Byrne, P. A., O'Reilly, S., and Quay, P. M.: 1979, 'Brain Death – An Opposing Viewpoint', *Journal of the American Medical Association* **242**, 1985–1990.
4. Charron, W. C.: 1975, 'Death: 'A Philosophical Perspective on the Legal Definitions', *Washington University Law Quarterly* **4**, 979–1008.
5. Engelhardt, H. T., Jr.: 1975, 'Defining Death: A Philosophical Problem for Medicine and Law', *Ann. Rev. Respiratory Dis.* **112**, 587–90.
6. Green, M. B. and Wikler, D.: 1980, 'Brain Death and Personal Identity', *Philosophy and Public Affairs* **9**(2), 105–133.

7. Haring, B.: 1973, *Medical Ethics*, Fides Publishing, Notre Dame, Indiana.
8. *In re Quinlan*. 70 N.J. 10, 355 A. 2d 647 (1976).
9. President's Commission for the Study of Ethical Problems in Medicine and Biomedical and Behavioral Research: 1981, *Defining Death: Medical, Legal and Ethical Issues in the Definition of Death*, U.S. Government Printing Office, Washington, D.C..
10. Tomlinson, T.: 1984, 'The Conservative Use of the Brain Death Criterion – A Critique', *Journal of Medicine and Philosophy* 9(4), 377–393.
11. Tooley, M.: 1972, 'Abortion and Infanticide', *Philosophy and Public Affairs* 2, 37–65.
12. Veatch, R. M.: 1978, 'The Definition of Death: Ethical, Philosophical, and Policy Confusion', in *Brain Death: Interrelated Medical and Social Issues*, edited by Julius Korein, The New York Academy of Sciences, New York, pp. 307–321.
13. Veatch, R. M.: 1975, 'The Whole-Brain-Oriented Concept of Death: An Outmoded Philosophical Formulation', *Journal of Thanatology* 3, 13–30.
14. Veatch, R. M.: 1976, *Death, Dying, and the Biological Revolution*, Yale University Press, New Haven, Connecticut.
15. Walker, A. E. *et al*.: 'An Appraisal of the Criteria of Cerebral Death – A Summary Statement', *Journal of the American Medical Association* 237, 982–986.
16. Warren, M. A.: 1973, 'On the Moral and Legal Status of Abortion', *The Monist* 57, 43–61.

RICHARD M. ZANER

BRAINS AND PERSONS: A CRITIQUE OF VEATCH'S VIEW

Professor Veatch's continuing argument with the President's Commission's recommendation of a "whole-brain" definition of death is important and bears brief rehearsal. As he pointed out, the Commission committed two basic errors. First, it mistakenly rejected the most plausible conception of death – that oriented toward the "higher brain functions" – in favor of a fundamentally "animalistic" view which gives "highest priority to the capacity to integrate bodily functions" ([6], p. 183), which is what the "whole-brain" definition really comes down to in the end.

The second mistake the Commission made concerns the demands of public policy: "it has missed the obvious and traditional way of resolving policy conflicts when matters of religious and philosophical variation are at stake," i.e, to insure freedom of choice on vital personal issues ([6], p. 185). For, on the one hand, the Commission's endorsement of the "whole-brain" conception is, like *any other* definition, an endorsement of a specific "theological/philosophical" viewpoint. On the other hand, to advocate the whole-brain conception as a matter of social policy is in fact *to impose* on people a specific religious/philosophical viewpoint – and it is one which is contrary to our Judeo-Christian heritage as well as to our traditional belief in the right of people to "exercise their consciences based on their (own) religious and philosophical beliefs . . ." ([6], p. 185)

Finally, in his review of the Commission's arguments for a whole-brain definition, Veatch contends that these arguments simply fail to make their intended point – either *against* the "higher-brain" concept or *for* the "whole-brain" notion. The mistake, as Veatch sees it, is to have reduced "higher-brain" oriented definitions to one that depends on the concept of "personhood". This move is not at all necessary; moreover, introducing this highly problematic notion merely clouds the issue. Hence, the Commission has not made the case it apparently intended to make, and in the course of its arguments it managed to violate both the prominent religious/philosophical viewpoint in our culture (the Judeo-

187

Richard M. Zaner (ed.), Death: Beyond Whole-Brain Criteria, pp. 187–197.

Christian view) and a major political belief (freedom of conscience and choice).

Veatch's alternative is twofold. First, he contends that "death should be the name we give to the condition under which it is considered appropriate in a society to initiate a series of (death) behaviors that are normally initiated when we call someone dead" ([6], p. 180). He nevertheless makes it very clear that he is quite unwilling to have that view imposed on other people. *Any* definition of death is implicitly an endorsement of one particular religious/philosophical view among other possible ones, including Veatch's. In the interests of preserving this pluralism, Veatch argues that "people should be permitted to examine their own religious and philosophical traditions and adopt the positions that are most plausible to them" ([6], p. 183).

Second, Veatch's position is that significant public policy issues enter in only because "picking a definition of death . . . has behavioral consequences" ([6], p. 183). And here, he believes, there is really only one area in which public policy mandates have any place: to limit possible extremist views. But this turns out to be far easier than might have been thought: for purposes of public policy, "we could adopt any one of the three plausible concepts of death referred to by the Commission with the proviso that individuals could opt for one of the others" ([6], p. 184).

I.

Veatch's point is clear: the Commission, "when it comes right down to it," never offers "a single argument that actually supports" its rejection of the "higher brain" function definition ([6], p. 173). What the Commission does, rather, is to reduce such "higher brain" definitions to one "depending on a concept of personhood" ([6], p. 174). This however, is not legitimate, for "there is no reason to assume that clarification of the concept of personhood must be achieved in order to consider someone dead based on loss of higher brain functions" ([6], p. 174). Indeed, so thorny and confused are these thickets that he seeks to avoid them entirely and to define death "without any reference to a theory of personhood" ([6], p. 175).

Although his article does, I think, strike the right emphasis on a more limited set of "higher brain" functions as appropriate for a definition of death, there is a quite puzzling feature about his analysis. He is clearly

anxious to "avoid" the "thicket" of "personhood" or "personal identity." This is puzzling for several reasons.

As was made manifestly clear in earlier papers [4, 7], there are few
more extraordinarily *personal* affairs in human life than dying and
death. After all, as Stuart Youngner and Edward T. Bartlett point out,
the first, most obvious question when someone dies is precisely: *who*
died?

On the other hand, Veatch himself seems hardly to have succeeded in
avoiding this "thicket." Objecting to the idea that he might be considered alive "simply because my body could gag," Veatch, accepting what
he takes to be the Judeo-Christian tradition, emphasizes two things:

First, I maintain that the human is fundamentally a social animal, a member of a human
community capable of interacting with other humans. Second, I maintain that I am *in
essence* the conjoining of soul and body – or to use the more modern language, mind and
body. If *either one* is irreversibly destroyed *so that the two* are irretrievably disjoined, *then
I* – this integrated entity – no longer exist. What is critical is the embodied capacity for
consciousness or social interaction. When this embodied capacity is gone, I am gone ([6],
p. 182), (emphasis added).

Now, while this brief indication surely falls short of being a "theory of
personhood," such a "theory" seems just as surely implicit here, and to
that extent Veatch's position is a clear case of precisely what the
Commission had in mind when it rejected "higher brain" formulations.
After all, it is what "I am in essence" that concerns Veatch, that is, an
embodied mind with capacities like "thinking, feeling, and interacting
with others" and this is what "essentially" constitutes a "person". Just
these "capacities", furthermore, are for him clearly dependent on, or at
least correlated with, those very "higher brain" functions which figure in
his own proposed definition. The fact that Veatch seems to have in mind
the ongoing *philosophical* dispute over "persons" and "personal" identity" (judging from the various citations in the body of his paper) – this,
I must say, is beside the point. His proper response to that dispute, if I
read him correctly, is by no means to pretend that he is, and has,
"avoided" this "thicket", but rather that his own version of this issue,
"personhood", is superior (for which claim there would presumably be
good reasons, of course).

To be sure, I (and others of us) would doubtless want to emphasize
the "disjoining" of what has hitherto been "conjoined" (supposing that
such terms can be given appropriate sense) as the core meaning of the
concept of human death. This being the *concept* at stake, Veatch's

analysis can then be seen to address mainly the *criteria* of death [*see* 7].
Even so, what obviously concerns Veatch is this "I," that "soul" or
"mind" and its concrete embodiment (mainly, by the "higher brain
functions") – and, as he mentions (and I have underscored), *either one*
of these may be destroyed. If destroyed (i.e., either the "soul" or its
"embodiment"), what was "conjoined" is now "disjoined". The stress
on "higher brain" *functions* thus does not in the least avoid necessary
reference to "person".

Veatch therefore does not succeed in defining death without reference to personhood: he accepts, rather, the *personal* concept of death,
and disputes the Commission's acceptance of a fundamentally *bodily*
concept of death. Along with that, he disputes the Commission's acceptance of "whole-brain" as the *criterion*, and opts for "higher brain
functions" as the proper *criteria*. Veatch's lack of success in so defining
death and delineating its criteria should neither surprise nor dismay us.
For if we were *unable* to answer the central question – "*whose* death?"
or "*who* has died?" – or were unwilling to do so, surely that would
rightly puzzle us, and it would be just as puzzling why a death can be and
so often is such a profound and moving, even awesome, experience,
especially when witnessed directly.

It would be equally puzzling, moreover, why so much feeling, suffering, resources, and the like are committed to keeping someone "alive",
even beyond the point at which we know our efforts are pointless. And
puzzling, too, why we so often persist in maintaining living bodies which
can hardly be called more than reminders, memorials, of that "someone" who was previously embodied but now is no longer, however much
respiration, circulation, reflexes, etc., persist. If it were merely "functions"
that had ceased, nothing of these things would be comprehensible.

Finally, unless we really were troubled and saddened by that "who"
which answers "who had died?", not even the President's Commission
Report makes much sense. In a sense, Veatch addresses the point here
when he rejects what he labels the "Lack of General Agreement"
argument.

II.

While the Commission's language is surely not that of feeling, passion,
and so on, but that of distanced deliberation appropriate for public
policy recommendations, it would nonetheless be quite pointless, even

as such, were the death being defined *not* the death of "this person" in "these" circumstances. And, in fact, I think that the Commission did in a way have just this kind of concern.

For instance, the *Report* is replete with words like "patient", "victim", "individual", and even "person" (even when it is not rejecting the "higher brain" formulation). As an illustration, when the Commission is considering various criticisms of "whole-brain" formulations, the *Report*, while supposedly talking about "the coordinated functioning of the various bodily systems," in fact talks about "patients" whose brains these happen to be ([3], p. 35). In this connection, it is then emphasized how different are "bodies lacking *all* brain functions and patients with intact brainstems" (*ibid.*) Here, and throughout this passage, the Commission's usage clearly suggests that this "difference" is in fact the presence or absence of the "person": *bodies* without any brain functions, and *patients* with only brainstem functions. If the former characterizes the "dead" and the latter the "living" ([3], p. 36), then "death" is marked precisely by the absence of the *person*.

Yet, looking into why the Commission rejected the "higher brain" definition, one thing is quite clear. Its main reason is not so much that these higher brain functions are not specifically correlated with such distinctive activities as "consciousness, thought, feeling", and the like ([3], p. 38). It is rather, as Veatch recognizes (pp. 174–176), that there is no general agreement on what "person" is:

crucial to the personhood argument is acceptance of one particular concept of those things that are essential to being a person, while there is no general agreement on this very fundamental point among philosophers, much less physicians or the general public ([3], p. 39).

Much the same thing is noted about the "personal identity" position.

The Commission then asserts the "slippery slope" worry about senile patients, the severely retarded, and so on, being regarded as "non-persons", and therefore "dead". Whatever one thinks of such "arguments", of course, their point should not be lost: "death" is the death of a "person", hence, a "non-person" would be just as "dead" as a "person" is "alive" – *however* "person" be understood.

Now, it seems unmistakeable in all this that the Commission was just as concerned in its way, as others of us, about death as a quite personal affair. Yet, while recognizing that the death of a person is connected with the irreversible cessation of "higher brain" functions, the Commission

curiously rejects that primarily because "personhood" seems so confusing and disputable a notion. It does not reject it because there are no
sound medical and biomedical evidences, and philosophical reasons, for
asserting a profound correlation between "higher brain functions" and
"person".

It is *this* oddity which should concern us, and not merely Veatch's
point that mere disagreement is no argument against accepting a position. Indeed, we should recall that Veatch himself tries to avoid what
the Commission rejected – "personhood" – and for much the same
reasons as those given by the Commission! Both Veatch and the Commission simply confuse issues here; while Veatch winds up in fact
endorsing some version of "personhood", as I've suggested, the Commission ends up endorsing a merely *bodily* view of death (what Veatch
calls "animalistic").

Both are wrong, and quite inconsistent with their respective analyses.
The source of both errors, finally, seems to be a confusion between the
concept of death, and the *criteria* by which death can be determined –
"whole-brain" versus "higher brain" functions. But since the Commission in particular seems to recognize that such capacities as thought,
feeling, and interacting with others are correlated specifically with those
"higher brain functions", however difficult it currently may be to *test* the
criteria determining personal death, the Commission, no less than
Veatch, seems to me committed to that concept of death. The dispute
concerns *criteria*, not the *concept*. In this respect, Veatch is quite right to
emphasize that the "higher-brain" formulation does not bring anything
new to the discussion, and that the "whole-brain" formulation re-
presents a radical break with traditional concepts. I conclude, then, that
what Puccetti, Youngner/Bartlett, and Veatch offer is the more correct
view, and that the Commission's view is one steeped in confusion and
because of that it draws the wrong conclusion.

III.

Several other remarks about Veatch's analysis may be made. In effect,
he argues that the most appropriate public policy is one that endorses
the "pluralism" of our times – i.e., allows people "to examine their own
religious and philosophical traditions and adopt the positions that are
most plausible to them" ([6], p. 183). There is, on the one hand, a
plurality of religious/philosophical traditions at hand nowadays; on the

other, to advocate a specific definition of death as a public policy is to endorse *one* specific religious/philosophical viewpoint. So far as it is a question of public policy, however, that endorsement is in fact an *imposition* of that religious/philosophical view on others, many of whom stand in quite different, often conflicting, traditions. There simply is no justification for this, and Veatch himself is clear that he is unwilling to impose his own religious/philosophical views on others. Likewise, neither the Commission nor the Federal government ought to impose any one particular religious/philosophical view on the general public. The only alternative, for Veatch, is a pluralistic public policy coupled with free choice: each of us should be encouraged to choose his own "definition" of death.

But I do not think that this will do. First, Veatch himself views *some* religious/philosophical views as unacceptable, incoherent, or wrong. Among these, of course, is the "animalistic" or "vitalistic" viewpoint of the Commission. As a good libertarian, Veatch wants to endorse the notion that people should be allowed to have their own respective views, and adhere to their implications so long as these do not unreasonably interfere with those of other people. But this effort strikes me as wrongheaded, for several reasons.

On the one hand, if it is true that any definition of death reflects some particular religious/philosophical view, it seems also true that each specific religious/philosophical view implies some particular public policy – or, more simply, has determinable social implications for the adherents' conduct. If it is *also* true that there is what MacIntyre calls "conceptual incommensurability" [1] among rival standpoints, then it surely follows that a heterogeneity of such religious/philosophical points of view will be reflected at the level of practical conduct. Rivalry and conflict at the one level will be reflected in rivalry and conflict at the other. That is, the endorsement of "pluralism" at the religious/philosophical level unavoidably leads to incommensurability at the practical level: interference with other people's views is also unavoidable, and, in Veatch's own terms, intolerable, as he maintained in his recent book on medical ethics theory ([5], pp. 110–114). Hence, Veatch's wish to permit people to choose their own definitions of death must, at the least, result in serious and perhaps irresolvable social conflict.

On the other hand, precisely to the extent that Veatch himself wishes to *argue* that some religious/philosophical views are incoherent, unac-

ceptable, or wrong, this would seem to imply that it would be quite
unreasonable for him to proceed to *argue* that people should be allowed
to choose such points of view. To make such a move seems to undercut
any clear sense to rational argument. What he perhaps should have
argued is that, for instance, the Commission's "animalistic" view is not
only one among many permissible public policies (among which people
could choose), but is also one among many theoretical views, each of
which is just as defensible as any other. This would, of course, be
dramatically inconsistent with the view he emphatically endorses in his
published book – that "we have to seek a universal or absolutist basis for
a medical ethics" on pains of vitiating the very idea of ethics itself ([5],
p. 113). It would, moreover, leave unanswered the severe problem of
social conflict among rival points of view. Whether or not ethics *must*
have a "universal or absolutist" foundation, and whether or not there is
some alternative to what Veatch in the present article argues about
"pluralism", – both are questions that cannot be pressed in this context.
I raise them only as issues that Veatch himself faces and does not
satisfactorily grasp, much less resolve.

A second reason why his proposal seems to me unacceptable arises
from the arguments and analyses presented by Puccetti and Youngner/
Bartlett: as they make exceedingly clear, not every view or claim about
the correlation between personal life and its neurological conditions is in
fact correct. Indeed, the evidence marshalled especially by Puccetti is
impressive, and suggests that some views are quite wrong – that of the
Commission especially. While the latter, as I've suggested, may arise
from a quite proper concern – the *concept* of death as the death of the
person, to be rigorously distinguished from both *criteria* and medical
tests – and does not in fact dispute the correlation in question, the
Commission wrongly ends up endorsing the whole-brain definition.

The implications for public policy seem just as clear: no public policy
which is incommensurate with what is known, medically and philosophically, to be true, can be acceptable.

Rather than argue for "free choice" of a "definition of death," then,
it would seem far more realistic, humane, and appropriate to urge that
choice be allowed only as regards the disposition of a person's *organism*
after his or her death. This, not choice of a definition of death, seems to
me far more consistent with the Anatomical Gift Act — contrary to
what Veatch suggests.

IV

A final remark should be made. Veatch takes his view to be within the standpoint of the "Judeo-Christian tradition". There may be reason to be dubious about this claim, however. He emphasizes here and elsewhere that it is "interaction with other people" or "being within a human community," which is fundamental for him: it is the *social* character of embodied human life that is "essential". We are given no analysis, of course, of just what constitutes this "social" character, or this "*interaction*" with other people. For that matter, neither are we provided any analysis of what constitutes the "conjoining" or "disjoining" of "mind" and "body", and thus are left very much in the dark about these clearly central notions (*see* [8], [9]).

That to the side, however, and more to the point here, neither does Veatch give much indication about what is specifically "Judeo-Christian" about his "essential" view. For the fact is that such emphases are by no means unique to that tradition. Beyond that, it has to be wondered whether Veatch's position really does stand within that tradition. A brief test from within his article can be used.

It is quite clear how Veatch wants to be treated, were he to be found with little more than "trivial bodily functions" intact: "I would want me considered dead in those circumstances" ([6], p. 181). It is decidedly quite *unclear* just how he proposes to consider *someone else* found in similar circumstances. Considering what he regards as the "one objection to the pluralistic solution" to the public issue, he mentions "the unconscious emergency room patient". This patient, he writes, "would be alive or dead depending on his views, and he would be in no condition to be asked what his views are" ([6], p. 184). The "simple solution" to this problem, Veatch says, is to give such people a choice – "dead or alive depending on his views". This, quite obviously, begs the question and gives no guidance whatever to those charged with deciding what to do in such cases (whether physician or family or both): such a patient, Veatch admits, is in no condition to be asked what he wants done, nor does he suggest any way to handle such cases. All we are clear about is (a) what *Veatch* wants done in his own case, and (b) that when a person indicates a choice about such matters we would be obliged to respect that choice. But this emergency room patient is left in a veritable limbo – like so many others these days. Nor would Veatch's "choice" proposal resolve all, or realistically very many, such cases in the future: he would *permit*, but not *require*, choice, and for any patient who did not choose, for whatever reason, the same limbo would be present.

And this is quite curious, especially for one who expressly wants to be within the Judeo-Christian tradition. That tradition, after all, is grounded in a profound sense of and continual effort to realize a human *community*, in which the *other person* is focal and prominent – and not, as in Veatch's argument, the self (i.e., Veatch). This sense of community, furthermore, as Richard Niebuhr emphasizes [2], is based on the prominence of *covenant* relationships. Presumably, Veatch would continue to endorse the version of "contract bargaining" he advocates in his work on medical ethics theory: "I prefer to recognize *covenant* as a kind of contractual relationship" ([5], p. 125). But if that is so, Veatch's position is *not* within the Judeo-Christian tradition at all, but in fact quite at odds with it. On the one hand, it seems to me quite indefensible to argue that covenants are but special kinds of contract. However that may be, to make them secondary to contract clearly runs contrary to that tradition.

What Veatch in fact endorses is the same Hobbesian individualism that is the centerpiece of his book. And the implications of that position concerning "interactions with others" are also quite clear: his world is one that, "in Hobbes's terms, is nasty, brutish, and short . . . [in which] one is to avoid the terror of a struggle of brute force pitting all against all . . ." ([5], p. 118). Not only, in such a "world", is it understandable that Veatch would be more concerned about his own case (the "false negative finding") than about those of others (the "false positive findings"). It is also clear that such a "world" is starkly in contrast with the "Judeo-Christian tradition" as I understand it. My point, of course, is not necessarily to endorse that tradition. It is rather to provide some assessment to Veatch's claim about it relative to his own position.

My criticisms of Veatch's position, it should be emphasized, do not in the least imply that the "higher brain functions" view is wrong. To the contrary, it seems to me precisely on target. Defending it, however, does not require the acceptance of Veatch's position – neither on public policy issues nor on the grounds of his professed Hobbesian commitments.

Vanderbilt University
Nashville, Tennessee

BIBLIOGRAPHY

1. MacIntyre, A: 1981, *After Virtue*, University of Notre Dame Press, Notre Dame, Indiana.
2. Niebuhr, H. R.: 1960, *Radical Monotheism and Western Culture*, Harper and Row Publishers, New York.
3. President's Commission for the Study of Ethical Problems in Medicine and Biomedical and Behavioral Research: 1981, *Defining Death: Medical, Legal and Ethical Issues in the Definition of Death*, U.S. Government Printing Office, Washington, D.C.
4. Puccetti, R: 1988, 'Neocortical Definitions of Death and Philosophical Concepts of Persons', this volume, pp. 75–90.
5. Veatch, R. M.: 1981, *A Theory of Medical Ethics*, Basic Books, Inc., Publishers, New York.
6. Veatch, R. M.: 1988, 'Whole-Brain, Neocortical, and Higher Brain Related Concepts', in this volume, pp. 171–186.
7. Youngner, S. and Bartlett, E. T.: 1988, 'Does Anyone Survive Neocortical Death?', in this volume, pp. 199–215.
8. Zaner, R. M.: 1964, *The Problem of Embodiment*, Phaenomenologica 17, Martinus Nijhoff, The Hague, The Netherlands.
9. Zaner, R. M.: 1982, *The Concept of Self*, Ohio University Press, Athens, Ohio.

EDWARD T. BARTLETT AND STUART J. YOUNGNER

HUMAN DEATH AND THE DESTRUCTION
OF THE NEOCORTEX

Up until the past two decades, there was little debate or discussion about definitions of human death. When a human being's heart stopped beating or his breathing failed, consciousness, internal integration, and the life of individual organs, tissues, and even cells ceased – quickly, inevitably, and permanently.

The rapid development of medical technology has recently given us the ability to support and even replace many of the body's vital organs and functions. In Intensive Care Units (ICUs), application of this technology has confused the traditional ways of distinguishing between the life and death of a human being. This confusion has stimulated a re-examination of, and lively debate over, competing definitions of death [15].

In the 1960s, this technology gave rise to a number of other practical problems – e.g., organ transplantation and the need for appropriate donors. There was also concern about the indiscriminate use of this new technology, without regard for patient and family wishes, human dignity, or cost. A discussion about the appropriate circumstances for stopping life-sustaining interventions was inevitable, as were the accompanying worries about euthanasia's slippery slope.

In 1968, an Ad Hoc Committee at the Harvard Medical School proposed "a new criterion for death": total and irreversible loss of functioning of the whole brain [1]. The publication of this report marks the beginning of the struggle to gain medical, legal, and societal acceptance for the concept of **whole-brain** death. The motivations for the report were clearly stated: 1. to relieve the "burden" imposed by severely brain-damaged patients; 2. to quell the "controversy in obtaining organs for transplantation" ([1], p. 337).

The report did describe tests that reliably demonstrated that the entire brain had permanently ceased to function. Although its title, "A Definition of Irreversible Coma," also suggests an interest in the concept of death, no attempt was made to answer the question, "What does it mean for a human being to die?" In fact, the report fails to distinguish

199

Richard M. Zaner (ed.), Death: Beyond Whole-Brain Criteria, pp. 199–215.
© 1988 *by Kluwer Academic Publishers.*

among, and often confuses, *definitions*, *criteria*, and *tests* of death. Since much of this confusion persists in the writings of the later whole-brain strategists, we will attempt to clarify these important distinctions.

I.

Any formulation of death has three components: a concept or *definition* of what it means to die; operational *criteria* for determining that death has occurred; specific medical *tests* showing whether or not the criteria have been fulfilled. Because they answer the question, "What does it mean for a human being to die?" definitions of death are conceptual – i.e., primarily abstract and philosophical. Criteria set the general physiologic standards for determining whether death, as defined conceptually, has occurred. Once criteria have been determined, specific medical tests can be developed to demonstrate their fulfillment.

Any adequate formulation (definition, criteria, and tests) of death must meet certain standards. First, specific criteria and tests must correspond to a given definition; otherwise the question, "Criteria or tests of what?" cannot be adequately answered. Second, the definition must be sound. It must accurately identify the quality that is so essentially significant to a living entity that its loss is termed "death".

It is senseless to fulfill accurately the criteria of an unsound definition. For example, if death were defined as the loss of sight, a medically sound diagnosis of permanent blindness would be considered proof of death. Despite the accuracy and reliability of the medical tests that demonstrate the permanent loss of sight, they cannot prove that the blind person is dead; the initial concept is wrong.

Conversely, a definition cannot be considered wrong simply because we lack specific, reliable tests to demonstrate the fulfillment of its criteria. The acceptability of a definition of death is strictly dependent upon how well it answers the question, "What does it mean for a human being to die?" Although specific methods for testing the fulfillment of its criteria may change and improve, a correct definition will remain the same.

Since criteria and tests serve to confirm the fulfillment of a definition or concept of death, their development should logically occur after the formulation of the conceptual question. Unfortunately, the whole-brain strategists have not followed this course. Until recently, they have largely ignored the need to have a definition at all.

Because the Ad Hoc Committee failed to deal directly with the central concept, it did not address the critical issue of including the entire brain in the criterion of death. The closest it gets to the identification of the essential brain functions is its discussion about the "burden" of those who meet the **whole-brain** criterion of death. It mentions only ". . . patients who suffer permanent loss of intellect . . ." ([1], p. 337). The Committee further notes that although modern technology can ". . . restore 'life' as judged by the ancient standards of persistent respiration and continuing heartbeat" to severely brain-damaged patients, such patients should, nevertheless, be declared dead, because " . . . there is not the remotest possibility of an individual recovering consciousness following massive brain damage" ([1], p. 337).

Furthermore, Henry Beecher, the Chairman of the Ad Hoc Committee, in an unpublished paper presented at the American Association for the Advancement of Science meeting in 1970, identified as important those functions supporting "the individual's personality, his conscious life, his uniqueness, his capacity for remembering, judging, reasoning, acting, enjoying, worrying, and so on" ([13], p. 311). This does not correspond to language adopted by the later **whole-brain** strategists. Terms like "the functioning of the organism as a whole," "the body's ability to organize and regulate itself," and the "integrating function" of the brain simply do not appear in the early **whole-brain** writings.

II.

There are two conceptually and anatomically distinct types of brain functions. The first type governs the central, neural integration of various vegetative functions – for example, spontaneous breathing; regulation of blood pressure; temperature regulation; neuroendocrine control. These integrative functions have been identified with the lower portions of the brain – specifically, but not exclusively, the brain stem. The second set of functions governs consciousness and cognition, and is linked anatomically to the cerebral hemispheres, particularly the neocortex. There are exceptions to these anatomical generalizations. Consciousness, for example, depends on the reticular activating system, which is in the brain stem.

Although the Ad Hoc Committee mentions only higher brain functions, later **whole-brain** strategists refer to both types of brain function as essential to a definition of human life and death. Thus, Veith *et al.* talk about ". . . some capacity to think, to perceive, to respond, and to regulate and integrate bodily functions . . ." ([14], p. 1653). They give no reason to choose between the ability to think and perceive and the ability to integrate body functions. Both are essential to their definition of human life. Yet, like Beecher and his Ad Hoc Committee, Veith does unmistakably favor the higher brain functions. Without the brain, he says, the residual activities do not confer an ". . . iota of humanity or personality . . . ." Without the brain, ". . . neither a human being nor a person any longer exists" ([14], p. 1653).

Korein spoke in a similar manner about the whole-brain as the criterion of death:

In the human organism, the brain is the critical component. Brain death then becomes equivalent to death of a person ([6], p. 19).

He goes on to say that:

If the brain is irreversibly destroyed, the critical system is destroyed, and despite all other systems being maintained by any manner whatsoever, the organism as an individual functioning entity no longer exists. If the critical system, i.e., the brain, in a man is destroyed, the human organism is no longer in a state of minimal entropy production; its state will progressively become more disorganized by spontaneous fluctuations. Therefore it will never return to its initial state as a sentient human being ([6], p. 26).

Like Beecher and Veith, Korein identifies a higher brain function, sentience, as essential. Why, then, does he include the entire brain (brain stem and neocortex) in his criterion? The answer has to do with a confusion between what death is and what causes it. For example, he discusses how the failure of the brain stem and its integrating functions leads to such an increasingly unstable situation that ". . . irreversible cardiac arrest will inevitably follow regardless of the maintenance of all resuscitative procedures" ([6], p. 27). In other words, persons whose brain stems are destroyed will soon "die", in spite of all heroic measures. Stated even more simply, destruction of the lower brain causes death.

Still, if sentience is really essential, why aren't people like Karen Quinlan, who remain in persistent vegetative states, considered dead? Korein's answer is pragmatic. "The determination of such irreversibility is still undergoing investigation, and prognostication is not always sharply defined" ([6], p. 27). In other words, completely reliable tests that determine just the irreversible loss of consciousness and cognition have not yet been developed. The whole-brain strategists insist on the inclusion of the lower brain in part because no one with whole-brain destruction has ever awakened. Fulfilling this criterion guarantees no false positive diagnoses of death. But, as we stated earlier, the technical problem of devising reliable tests is not a good reason for questioning the adequacy of a definition.

III.

One must also contend with the spectre of the slippery slope. Acceptance of a definition of death based only on consciousness and cognition could lead to the unjustified killing of the senile and the retarded [2], [5], [8], [12]. It would be extremely distasteful to "suffocate" or "bury" spontaneously breathing, but permanently unconscious patients ([2], p. 254). This argument is used repeatedly by the whole-brain strategists.

At this stage, the whole-brain theorists did not have a clear concept on which to base their whole-brain criterion of death. Some commentators stated that controversy over this issue should be avoided in favor of solving the ". . . practical problems facing physicians, families, and the community . . . ." The whole-brain criterion would hopefully be ". . . acceptable to persons holding various concepts of death . . ." ([11], p. 49). In fact, public and legal acceptance of the whole-brain criterion had proceeded apace, in spite of this conceptual vacuum. The utilitarian arguments against a definition based on consciousness alone and the need to expedite organ transplantation were quite persuasive to legislatures, policymakers, and the medical community. By 1981, The American Medical Association and the American Bar Association had both endorsed The Uniform Determination of Death Act; twenty-seven states had enacted legislation recognizing a whole-brain criterion of death ([8], pp. 67, 119).

The report of the Ad Hoc Committee marked the beginning of the struggle to gain medical, legal, and societal recognition for brain death. The 1981 publication of the President's Commission report "Defining Death" signifies an end to that era by recommending that all U.S. jurisdictions accept the Uniform Determination of Death Act.

In spite of this endorsement, the President's Commission perpetuates the whole-brain supporters' bias against theoretical issues. For example, the Commission disavows the need "to resolve all the differences among the leading concepts of death," and goes on to say that ". . . philosophical refinement beyond a certain point may not be necessary . . . for the purpose of public policy . . ." ([8], p. 36). While not taking a clear stand on the issue, the President's Commission certainly seems to favor the definition of death presented by Bernat and his associates in 1981. Their article, "On the Definition and Criterion of Death," which appeared in the *Annals of Internal Medicine*, is the first concerted effort to discuss the conceptual issues underlying the choice of a whole-brain criterion [2]. Unlike its predecessors, it offers a philosophic argument in support of this choice, and specifically rejects any definition based on higher brain functions. After 13 years and considerable political success, the whole-brain strategists were finally developing a conceptual framework. Let us evaluate its success.

IV.

Up to this point, the whole-brain advocates had mentioned both integrative and cognitive brain functions. Although no direct choice was made, they seemed to favor the higher brain functions. Bernat *et al.*'s definition of death as the permanent cessation of "functioning of the organism as a whole" is the first clear move away from this position [2]. The most startling and ultimately indefensible feature of their definition is the failure to describe both types of brain function in elaborating on the meaning of "function as a whole". They focus exclusively on the brain's capacity to regulate and integrate vital, but distinctly reflex, vegetative functions. Nowhere do we see the terms "intellect", "sentience", or "humanness", which were so dear to the earlier whole-brain strategists. In fact, Bernat *et al.* assert that the term "person" is too "vague", and that death can only be applied to organisms, not persons ([2], p. 390).

In a more recent publication, Bernat has added "awareness of self and environment" to his original list of vegetative functions; no explanation or comments are offered ([3], p. 47). The clear implication is that awareness of self and environment is just one of many equal items in the smorgasbord of brain functions.

In order to evaluate this notion as well as other aspects of the whole-brain definition, we will consider three cases involving patients who have suffered significant and irreversible brain damage.

Case 1: The classic brain-dead patient: destruction of the entire brain. This patient has suffered massive destruction of the entire brain, resulting in the loss of both higher and lower functions. In addition to being permanently unconscious, he has lost all ability to organize vital vegetative subsystems, such as breathing, neuroendocrine control, blood pressure, temperature regulation, etc. However, many of these functions can be supported or replaced by skilled personnel and sophisticated technology. After the patient is put on a respirator, the staff monitors his blood oxygen concentration and adjusts the rate and depth of breathing. Blood pressure is checked and supported, if necessary, with vasopressors. Electrolyte and fluid balances are monitored and adjusted when necessary. Nutrition is provided intravenously through a hyperalimentation line.

The patient's heart continues to pump blood, which has been oxygenated and loaded with nutrients, to various parts of the body, enabling the continuation of other functions. The skin is warm, the kidneys continue to produce urine, hair grows, and the liver continues its many functions, such as the removal of waste products from the blood. This patient is a classic example of the destruction of the whole-brain.

Case 2: The Karen Quinlan-type patient: destruction of only the neocortex. Unlike the preceding case, the patient's lower brain structures have been left intact and functioning. Although he can breathe spontaneously and maintain temperature and blood pressure, he has irreversibly lost the functions of consciousness and cognition. Although such a patient requires much less attention and technical support than the first one, he would not survive long without assistance. Liquified food must be placed directly into the stomach through a gastrostomy or nasogastric tube. Because he is incapable of moving on his own, he must be frequently turned to prevent the development of bedsores, which would lead to infection and "death". His excretions must also be managed. The patient is unable to do or understand anything at all.

Case 3: The locked-in patient: destruction of only the lower brain. This patient suffers significant, but incomplete, destruction of the brain. Unlike the Quinlan-like patient in Case #2, the brain portions responsible for consciousness and cognition are intact. All portions of the brain stem and deep cerebral areas responsible for integration of vegetative functions have been destroyed. However, the blood supply and neural connections to the other cerebral areas, as well as the reticular activating system located in the brain stem, have been spared. Although the patient cannot spontaneously regulate respiration, blood pressure, temperature, hormonal balance, and other functions, he is awake and alert. Let us also assume that the brain stem areas responsible for hearing and eye movement remain unaffected, so that he can give meaningful responses to questions by moving and blinking his eyes (e.g., one blink means "yes", and two blinks mean "no"). Like patient #1 with total brain destruction, this patient's life can only be maintained through the full efforts of the Intensive Care Unit staff.

Let us first examine the similarities among all three types of patients. According to the traditional definition and criteria of death, all three patients would clearly be considered alive. In all three cases, the hearts are beating and oxygenated blood is being circulated throughout the body – i.e., vital fluid flow persists.

Without continued support and intervention, life, by any definition, would soon come to an end in all three cases. Within days of the withdrawal of feeding and fluids, the patient in a persistent vegetative state would experience circulatory collapse, cardiac arrest, brain stem failure, and death of individual organ systems, organs, tissue, and cells. In the other two cases, withdrawal of support would lead to a similar total destruction within minutes.

Patient #1 with no brain functioning would be considered dead by the whole-brain criterion. Patients #2 and #3 would be judged alive because some brain function remained. We must not forget, however, that the whole-brain criterion is unacceptable unless it operationalizes a definition of death that logically requires the functions of the entire brain. An analysis of cases #2 and #3 reveals that no such definition is forthcoming from Bernat and his associates or the President's Commission.

Proponents of the various brain-oriented formulations of death would agree that some sort of life persists in all three of these patients. The lives of individual cells, tissues, organs, or even organ systems may persist after the human being or human organism's life has ended. In fact, the successful retrieval from dead donors of living organs for transplantation was a fundamental stimulus for the reexamination of human death.

If we are to reject the traditional heart-lung, vital fluid flow formulation of death, but are also unwilling to accept one that requires absolutely no metabolic activity whatsoever, we need a method for choosing among the brain-oriented formulations illustrated by the three cases.

The arguments of Bernat *et al.* provide no guidance in developing a standard on which to make this choice. This leads to two major problems: 1. Expressions like "as a whole" are unclear, vague, and indefinite. 2. Bernat's arguments characterize the death of the wrong type of organism – i.e., one that is not a human being.

The whole-brain criterion proposed in Bernat *et al.*'s original paper did not correspond to their definition, which focused on the brain's non-cognitive integrating functions [2]. All that is needed to fulfill such a definition is the destruction of the brain stem, not the whole-brain. Since cases #1 and #3, the patients with whole-brain and only lower brain destruction, respectively, have lost the ability to integrate the body's subsystems, they would have to be considered dead by Bernat's *definition*. According to their *criterion*, however, only patient #1 is dead; he is the only one whose entire brain has been destroyed.

There are two ways to resolve this inconsistency. First, the criterion of death could be narrowed to fit the definition – i.e., only include those brain areas connected with non-cognitive regulation and integration. This change would, however, make it necessary to consider patient #3 dead; this is a difficult thing to do when he can give meaningful answers to questions like, "Do you feel pain?" and, "Are you a baseball fan?"

In a real sense, Bernat *et al.* have simply reversed the mistake of the earlier brain strategists, who also picked a whole-brain criterion of death but emphasized only the higher brain functions (intellect, sentience, personality). This type of problem arises when you try to develop a definition of death from a criterion, instead of vice versa.

The second, and more reasonable, choice for Bernat and his colleagues is to expand the definition to fit the criterion. This entails the addition of the concepts of consciousness and cognition to the meaning of the expression "function as a whole". As we mentioned earlier, Bernat himself seems to have made this choice. Three years after their original paper, he quietly adds "awareness of self and environment" ([3], p. 47) to a list that includes respiration, circulation, and integration of organ systems. Is this enough? Does the concept of "functioning as a whole", even with the addition of higher brain functions, adequately describe what is essential to the life and death of a human being? Although this is a step in the right direction, it doesn't solve the problem. Because there is still no standard, there is no way to determine which of the various subsystems is the most important. Bernat *et al.* have argued that death is an event. How can we tell when it has occurred?

According to Bernat's most recent argument, the Quinlan-type patient in case #2 has lost only one attribute, "awareness of self and environment." Because he ". . . maintains the majority of functions of the organism as a whole . . . he is clearly alive" ([3], p. 48). This makes it a numbers game; life or death is determined by counting the number of remaining attributes. Such a quantitative standard is acceptable only if all of the attributes are equally important in establishing that someone is alive. Let us see if this works out. Bernat *et al.* observe that, ". . . when a higher organism is comatose, functioning of the organism as a whole may still be evident, such as temperature regulation . . ." ([2], p. 390).

So, if a human being has lost the attribute of consciousness and cognition, but retains the ability to control his temperature, he is still alive. Suppose he has a small stroke in the area of the lower brain responsible for temperature regulation and permanently loses that "attribute" as well. Two attributes are now gone. Is he dead yet? What if his temperature is regulated by sophisticated machines and medical personnel in an ICU? Is he still functioning "as a whole"? Since two are hardly a majority, it would seem that he ought to lose more before he is pronounced dead. Suppose he has another stroke that damages his respiratory and circulatory centers. But skilled nurses and physicians place him on a respirator, monitor his blood gases, and appropriately adjust machine settings and oxygen levels. Blood pressure is maintained by the use of vasopressors. The patient has lost consciousness, as well as the ability to integrate and regulate temperature, respiration, and blood pressure. Does the loss of four attributes render him dead? Has he lost a "majority" or the appropriate "cluster" ([3], p. 48; [8], p. 36) of attributes? He still has neuroendocrine control and electrolyte balance. What if he loses these and they are restored by an ICU staff?

In fact, the idea of counting attributes as though each one were a discrete occurrence, readily identifiable and distinct from other events, needs to be carefully examined. It makes sense to talk about a majority only if there is agreement on the total number of attributes.

Even after adding the higher brain function "awareness of self and environment" to the list of vegetative functions, Bernat's account fails. The problem remains the total lack of any qualitative standard whatsoever, not just quantitative uncertainty. If, for example, we really were to make decisions on the basis of ". . . a majority of the functions of the organism as a whole . . ." ([3], p. 48), patient #3, who retains only higher brain function, would be considered dead; he has lost a majority of those functions. To pronounce him dead when he is awake and responsive is obviously untenable. Do we really want to say that a patient whose only remaining brain function is temperature regulation is alive (or dead) in exactly the same sense as the patient whose only attribute is consciousness and cognition?

Obviously, the solution to the problem requires more than quantitative precision. Some attributes are more important than others. In order to determine which attributes are more important than others, we need a standard of selection. To the best of our knowledge, no such standard exists in any of the arguments of the whole-brain theorists.

The absence of such a standard gives rise to an even more fundamental problem: Bernat *et al.* end up describing the death of something that is not a human being. In this regard, they caution that, "We must not confuse the death of an organism which was a person, with an organism's ceasing to be a person" and that ". . . death can be applied directly only to biological organisms and not to persons . . ." ([2], pp. 390, 391).

These remarks add up to the following argument. Persons are not biological entities. The concept of death is appropriately applied only to the latter. *Therefore*, persons cannot properly be said to die. We believe this argument is unsound, primarily because the first premise is false. One cannot understand what it is to be a person if one adopts a Cartesian view – i.e., excluding material predicates [10]. While there are non-biological dimensions to personhood, its functions are as surely grounded in the biology of the central nervous system as are the abilities to regulate temperature and blood pressure. Nevertheless, if the presence or absence of personhood is to be excluded from the determination of death, the supporting physiology must also be excluded. The physiology of death would, therefore, be restricted to the brain stem functions essential to Bernat's "human organism". Considering case #3 from this standpoint results in some rather strange consequences. Since this patient has lost all brain functions except those associated with personhood, we must conclude that Bernat's human organism is dead. However, the patient still blinks meaningful answers to our questions about his welfare. We are thus communicating with a person who according to Bernat is neither dead nor alive, but whose human organism has died.

Bernat himself seems confused on the issue. At one point he says that "Personhood is a psychosocial rather than a biological concept and refers to certain abilities and qualities of awareness of self and environment" ([3], p. 47). Yet, shortly afterwards, in describing his revised explanation of "the functioning of the organism as a whole," he says:

In more *physiologic* terms, this concept refers to the vital functions of respiration and circulation, awareness of self and environment, the integration of organ systems through neural and neuroendocrine control . . . ([3], p. 47; emphasis added).

V

In criticizing *our* attempt to develop a higher brain account, Bernat *et al.* say that we ". . . fail to make the critical distinction between loss of personhood and death" ([4], p. 456). It is true that we do not make this distinction. The point of our argument is, in fact, that the permanent loss of personhood because of the irreversible destruction of the cortex is death. We make no distinction because we believe there is no distinction to be made.

We do not wish to get into an involved discussion of the philosophical problem of personhood, especially as it shades into that of personal identity. We believe, however, that Bernat's criticism is both important and answerable without an undue amount of philosophical paraphernalia. It is confused and confusing to talk about the loss of personhood as though it were an attribute. To speak of the loss of personhood as though it were akin to the loss of childhood is to speak of it as though it were just another transient condition or quality. The loss of childhood is followed by entry into another phase of life – e.g., womanhood; this involves an exchange of predicates attributable to the same continuous subject. The loss of personhood involves the elimination of the subject. To make the point in a Kantian manner, "person", like "existence", is not a predicate. Indeed, person is a subject, to which appropriate predicates may be attached. Loss of personhood cannot be understood as the loss of an attribute, which could be followed by death. Once it has been lost, the subsequent death is not that of a human being; instead, it is the death of a thing. The issue thus becomes one of choosing the type of entity with whose death we are concerned.

We do not deny that life may go on after the person has died – just as Bernat cannot deny that it may continue after the whole-brain has been destroyed. It is a question of when we no longer have a living human being. We are not concerned with the death of the organism that outlives the person.

If we focus only on the demise of a breathing body that has outlived its owner, and if we seriously argue, as does Bernat, that death applies only to this body, then we simply cannot explain the whole cluster of feelings and thoughts that ordinarily surrounds death. When death has occurred, the first thing any ordinary person wants to know is who died. Death is an extraordinarily personal event. We grieve the loss of a loved one in a manner that would not be appropriate for the loss of a breathing body or a cluster of organs. We would not be consoled to learn that even though a loved one had died, a number of organs were continuing to function. To suggest that man's age-old fascination with, and fear of, death has just been a concern over the status of organs and organ systems is to make a bad joke. We believe that using the word "death" to describe the demise of a mindless organism is a corruption of its ordinary meaning. Death is not the loss of something to someone; it is the loss of someone. Because a human being is a person, the irreversible destruction of the neocortex — i.e., the center of consciousness and cognition, constitutes death.

VI

The issues of "life" and "death" are confused by the fact that "life" can exist at different levels of organization. Cells, tissues, individual organs, and even organ systems can all be said to be alive. For example, human cells can be grown in tissue culture; a "live" vein from the leg can be removed and transplanted into the wall of the heart; "live" heart and lungs can be transplanted into another person's body where they continue to oxygenate and circulate the blood.

In all three of the cases presented earlier, individual cells, tissues, and **organs** continued to live, requiring various degrees of external integration and control to remain functional. But we do not say that a human being is alive simply because some of his cells are growing in a petri dish or his kidney is functioning in another person's body. When Bernat and the President's Commission consider the patient in case #1 dead, in spite of the continuation of "life" at the organ, tissue, and cellular level, they reject the "metabolic activity" theory of life. A patient with total brain destruction, like case #1, is, indeed, dead, in spite of this residual activity. But, having rejected this sort of "life" in case #1, they accept it in case #2, the Quinlan-type patient. In fact, we and other higher brain advocates are criticized for confusing death with the loss of personhood, since "life" obviously continues after personhood has ceased ([4], p. 456). This "life" that lingers on after the destruction of the neocortex is metabolic activity that is only differentiated by complexity. At most, the complexity is that of a mindless organism; while it is a human organism, it is no more a living human being than a human kidney or human cell functioning in vitro.

It is certainly possible to speak of consciousness and cognition in terms of metabolic activity. It is also possible to differentiate between that kind of metabolic activity and the metabolic activity of Bernat's "human organism". The activities supported by the metabolic processes of the neocortex (consciousness and cognition) are essential to a proper understanding of the concept "human being"; the corresponding activities of the "human organism" are not. These activities lack conceptual importance, and even their medical significance is diminishing.

VII

Because we place such importance on the concept of "human being" as the standard of acceptability, Bernat and his colleagues have accused us of a ". . . misleading emphasis on human death . . . ." They reject the higher brain formulation by denying ". . . that there is any special sense of death that applies only to humans" ([4], p. 456).

Mayo and Wikler express the same concern more generally:

Humans are only one kind of living thing: what is essential to being human is something that differentiates us from other living things, something which we do not share with dogs or trees or mosquitoes. Being alive, on the other hand, is something we do share with other living things ([7], p. 209).

In fact, all three cases (the classic whole-brain dead patient, the **Quinlan-type** patient, and the patient with only cerebral functions) do share something with dogs and mosquitoes – i.e., the life of individual cells, tissues, organs, and organ systems. Yet, for Bernat and the President's Commission, such "life" is not enough; they claim that case #1, with whole-brain destruction, is, indeed, dead.

Presumably, then, the common factor that defines the lives of dogs, mosquitoes, and men is the integrating capacity of the central nervous system. Accordingly, the Quinlan-type patient in case #2 is alive because he shares this capacity with insects and canines. Case #3 retains only the one brain function we believe essential to human life – consciousness and cognition. It would be interesting to inform patient #3 that he was now dead because he had lost what he used to have in common with dogs and mosquitoes. His responses might be even more interesting.

Moreover, Bernat and his colleagues confuse what is essential and what is unique. We have never claimed that death either did or did not involve the loss of something uniquely human. We did say that the death of a human being was the absence of that which was essential to his life. Whether or not such a feature is unique is not a logical consequence of its being necessary. Uniqueness is an empirical consequence of the facts; necessity is a conceptual matter. Only the latter is appropriate in defining human death. The higher brain functions of human beings are, to some extent, possessed by other mammals – e.g., consciousness; rudimentary cognitive skills; feelings.

VIII

Bernat and his colleagues use a variety of philosophical arguments involving language, language use, and social convention to support their own view and to attack the higher-brain theory. In spite of their populist ring, these arguments have the effect of stifling dissent before it can be heard.

They characterize their position as making ". . . explicit the ordinary meaning of the term 'death' implicit in everyone's usage of the word" ([4], p. 456). They go on to argue that since "death" is a non-technical word, we and other advocates of a higher-brain definition are wrong because we use the word differently than Bernat, his colleagues, "and nearly everyone else." Finally, and more substantially, they say that our argument ". . . confuses knowing the meaning of the word 'death' with knowing facts about that event to which the word 'death' refers" ([4], p. 456).

The importance of knowing empirical facts is determined by the types of words in question – i.e., specialized and technical vs. ordinary. It would, indeed, be odd to argue that the "real" meaning of ordinary words like "try" or "intend" was radically different from the ways in which they were normally used. But, at the other end of the spectrum, there are non-ordinary words whose meanings are entirely dependent on special knowledge. Eventually, words like "neutrino" and "papilloma" may work their way into ordinary language; they do not, however, begin there. To ask, "What does the word 'neutrino' or 'papilloma' mean?" is the same as asking, "What *is* a neutrino or a papilloma?" To understand the meaning of the word "neutrino", one must know something about sub-atomic particle theory. The dictionary definition of "neutrino" is: "a neutral particle smaller than a neutron, having a mass approaching zero" ([15], p. 986). In providing the meanings of this word, this dictionary draws our attention to some facts – i.e., it tells us what a neutrino is. The definition is preceded by the qualification "in physics", which makes it clear that "neutrino" is not to be thought of as a common word in ordinary language. Because its meaning is so closely related to the facts of this science, it is considered a technical, rather than an ordinary, term.

Unlike "neutrino", the word "death" certainly belongs in the vocabulary of any competent, native speaker of English. The fact that it is a common word, and not the special creation of a special science, does not, however, mean that its meaning is determined in the same manner as that of the verb "to try". Like the concepts of "illness" or "intelligence", an understanding of "death" requires some special knowledge about human physiology and anatomy. In fact, it is only because of medicine's recently acquired ability to understand and control the biology of dying that we are faced with the current problem of determining the correct meaning of "death". If we look up the word "death" in a medical dictionary, we are confronted with a variety of factual claims. For example:

In man, death is manifested by the loss of heart beat, by the absence of spontaneous breathing, and by cerebral death ([9]).

Indeed, the question of whether death is manifested by the loss of heart beat constitutes the factual issue that has forced medicine to reexamine the concept.

Generally speaking, the difference between facts and language is not as clear as Bernat would have us believe. Particularly with respect to "death", matters of fact and matters of language are, and should be, intertwined. Were it simply a question of determining how people commonly spoke, it would not have stirred up such interest and confusion in both the medical and philosophical communities. This confusion over correct usage has prompted England, Canada, Australia, and the United States to appoint commissions to address the issue formally. Under such conditions, it is extraordinarily ill advised to resolve things according to ". . . the way everyone uses the word" ([4], p. 456).

Bernat and his colleagues reject our view by saying that if a higher-brain definition were accepted, ". . . people would be declared dead who are considered living by the law, culture, and tradition of *every country in the world*" ([4], p. 456; emphasis added). If that is true, the odds against us are, indeed, incredible. But, if support of this sort is needed before a concept can be considered correct, the whole-brain theorists themselves were not correct until 1981, when Colorado and Idaho became the 26th and 27th states to enact legislation recognizing a whole-brain criterion of death. While current thinking, speech, laws, and customs need to be taken into account, one must not reject a definition merely because it challenges the status quo. The whole-brain strategists did not take such a reactionary position when they first expressed their own views in the 1960s.

As indicated by the endorsement of commissions and legislatures, public acceptance is the result of a complex and lengthy process. One would certainly hope that one of the determining factors is the soundness of the arguments supporting the view in question. However, it would be incredibly naive to think that only those positions supported by sound arguments ever won acceptance. Acceptance by a national commission or state legislature is as dependent on practical and political accommodation as it is on sound reasoning.

We believe that only the higher brain functions, consciousness and cognition, define the life and death of a human being. If the remaining philosophical problems with such a definition cannot be resolved, we must return to a metabolic activity or vital fluid flow definition of death. The whole-brain theories simply do not work.

Cleveland State University
        and
Case Western Reserve University
Cleveland, Ohio

BIBLIOGRAPHY

1. Ad Hoc Committee of the Harvard Medical School to Examine the Definition of Brain Death: 1968, 'A Definition of Irreversible Coma', *Journal of the American Medical Association* **205**, 337–340.
2. Bernat, J. L. *et al.*: 1981, 'On the Definition and Criterion of Death', *Annals of Internal Medicine* **94**, 389–394.
3. Bernat, J. L.: 1984, 'The Definition, Criterion, and States of Death', *Seminars in Neurology* **4**, 45–51.
4. Bernat, J. L. *et al.*: 1984, 'Definition of Death (letter)', *Annals of Internal Medicine* **100**, 456.
5. Horan, D. J.: 1978, 'Euthanasia and Brain Death: Ethical and Legal Considerations', *Annals of the New York Academy of Sciences* **315**, 363–375.
6. Korein, J.: 1978, 'The Problem of Brain Death: Development and History', *Annals of the New York Academy of Sciences* **315**, 19–39.
7. Mayo, D. and Wikler, D. I.: 1979, 'Euthanasia and the Transition from Life to Death', in W. L. Robinson and M. S. Pritchard (eds), *Medical Responsibility: Paternalism, Informed Consent, and Euthanasia*, Humana Press, Clifton, New Jersey, pp. 195–211.
8. President's Commission for the Study of Ethical Problems in Medicine and Biomedical and Behavioral Research: 1981, *Defining Death: Medical, Legal, and Ethical Issues in the Determination of Death*, U.S. Government Printing Office. Washington, D.C.
9. *Stedman's Medical Dictionary*: 1983, 24th edition, Williams and Wilkins, Baltimore, p. 346.
10. Strawson, P. F.: 1959, *Individuals*, Methuen & Co Ltd, London, Pt. I, sec. 3, 'Persons', pp. 81–113.
11. Task Force on Death and Dying of the Institute of Society, Ethics, and the Life Sciences: 1972, 'Refinements in Criteria for the Determination of Death: An Appraisal', *Journal of the American Medical Association* **221**, 48–53.
12. Tendler, M. D.: 1978, 'Cessation of Brain Function: Ethical Implications in Terminal Care and Organ Transplant', *Annals of the New York Academy of Sciences* **315**, 394–397.
13. Veatch, R. M.: 1978, 'The Definition of Death: Ethical, Philosophical, and Policy Confusion', *Annals of the New York Academy of Sciences* **315**, 307–321.
14. Veith, F. J. *et al.*: 1977, 'Brain Death: I. A Status Report of Medical and Ethical Considerations', *Journal of the American Medical Association* **238**, 1651–1655.
15. *Webster's New World Dictionary*: 1964, College Edition: The World Publishing Company, Cleveland, Ohio, p. 986.
16. Youngner, S. J. and Bartlett, E. T.: 1983, 'Human Death and High Technology: The Failure of the Whole-Brain Formulations', *Annals of Internal Medicine* **99**, 252–258.

PART IV

THE CULTURAL CONTEXT

MARX W. WARTOFSKY

BEYOND A WHOLE-BRAIN DEFINITION OF DEATH: RECONSIDERING THE METAPHYSICS OF DEATH

I am going to argue here against a whole-brain definition of death, *and* against a neocortical definition of death, or against any other clinical *definition* of death, on two grounds: *first*, that none of these is, in effect, a *definition* of death, but are rather elliptical expressions for basic definitions of death, and hide more presuppositions than they reveal. In themselves, these presumed definitions are shorthand expressions for the underlying biological processes or functions that are taken to be the necessary and sufficient conditions for a human life – or what one may call a life-identity. They are at best discussion-openers, not definitions, and they sit somewhere between a definition of death, and the criteria for determining when that definition has been satisfied.

Second, I will argue for a social, and even an historical definition of human death – not on the grounds that death is *not* a biological fact, but on the grounds that calling it a biological fact is misleading as to what *sort* of biological fact it is. The argument here would be (though I don't develop it here) that *human* biology is the biology of essentially social-historical beings – it is *funny* biology, therefore not to be confused with or reduced to animal biology, though its precondition is animal biology, as animal death is the evolutionary precondition for human death. (I am talking emergent-levels talk here, in case it isn't obvious.) Further, I am going to argue against the Cartesianism – good, old-fashioned metaphysical Cartesianism – which it seems to me dominates all discussions of consciousness as the individuating, criterial feature of human life and personal identity, and thus dominates discussions of what constitutes the death of a human individual, or of a person. Again, this is not because I *don't* take consciousness, and even self-consciousness, to be the normal hallmark of human identity, and hence of *a* human life, but rather because the Cartesian version of this consciousness situates it in an *I*, which operates as a self-sufficient, asocial, and ahistorical entity.

Finally, I'm going to argue that the question of the definition of death is a question concerning the peculiar human requirement that a person has to be *declared* dead, i.e., that human death, unlike animal death, is a socially constituted fact requiring a judgment; and that to understand

Richard M. Zaner (ed.), Death: Beyond Whole-Brain Criteria, pp. 219–228.
© 1988 *by Kluwer Academic Publishers.*

the question of the definition of death properly is to understand how and why the locus of this judgment has traditionally been placed in the medical profession.

In order not to mislead any more than is useful, let me say categorically that I am not out to deconstruct death. What I will offer is in the way of a somewhat Hegelian, somewhat Feuerbachian, somewhat Marxist, and somewhat way-out and weird approach. The title of the essay is not my own. Yet, the suggested metaphysical orientation suits me: for what is at issue is a metaphysical question – or as I might prefer to say, an ontological one – and also an epistemological one: namely, what kind of being is human being, and what kind of fact about that being is its death? And how is it possible to come to know what this fact is? My essay is funnel-shaped. I begin wide and end narrow. What isn't clear is whether everything I'm going to pour into the wide end will come out at the narrow end, so I ask your forbearance. Now to begin.

I.

Spinoza writes, in the *Ethics*, that philosophy should be a meditation on life, not on death. He also writes, however, that every determination is a negation: *omne determinatio negatio est*. That is to say (if we put these separate assertions together), that to determine what life is, is to determine also what is not-life. The dialectic of life and death – whether we take it as a conceptual dialectic of interdefining terms, or as an ontological dialectic of "opposites", i.e., of mutually exclusive modes of being – is as central a theme in philosophy as it is in human life – and for the simple reason that both life and death are fundamental facts of our awareness. We are, as far as we can tell, the only species that is aware of its mortality, or for which life is a conscious, or self-conscious, value. All animals live and die. But part of what distinctively marks the human species *as human* is that the fact of life and the fact of death are constituted facts of our awareness, and that we constitute ourselves as being the kinds of beings we are, in one fundamental respect, by our self-appreciation as living and by the recognition of our mortality. We are that species, the individuals of which know themselves to be what they are in and through their characteristic life activities, and in their mutual (i.e., social) recognition of each other *as* human, i.e., as self-conscious, valuing, and socially interdependent beings. As we come to know ourselves, and others, and ourselves through these others that we take to be like ourselves, through the forms of life activity that are characteristic of the human – through language, through work as the production of the means of our existence, through love, or hate, or solidarity, or competition, through the reproduction of species life from one generation to the next – so too, we come to know ourselves, or *become* who we are, through the various social, cognitive, affective forms in which we acknowledge and declare the fact of death, not as an abstract fact, or as a concept only, but as a concrete-individual fact. Only individuals live, and only individuals die. Therefore, the recognition of the fact of an individual death, and the social declaration of this fact – the mode of asserting that an individual death has occurred, or of making the social judgment that an individual is dead, is itself a matter of how we define ourselves, constitute ourselves as human, create or transform our humanity. It is no small matter. But what sort of a large matter is it?

Now, I must stop and explain why I choose to begin with so grand an approach, for surely, in considering the question of the whole-brain definition of death, there is a more focused issue to examine: on what grounds can we determine whether the effective and irrevocable cessation of whole-brain activity *defines* the death of an individual?

II.

To begin with, as I suggested at the outset, the question is ill-formed and, on reflection, I withdraw the term "whole-brain *definition* of death" from my title. (I feel easy about that, since the title isn't my own.) For neither "whole-brain", nor "cortical", nor "vital fluid", nor "heart and lung" can serve as *definitions* of death, but rather only as criteria for determining when any given definition of death has been instantiated. We need instead, as many authors here have recognized, and as much of the literature does as well, a definition of death from which it would *follow* that the irrevocable loss of whole-brain or neocortical or *other* function *would* serve as criteria. So I want to begin again with the question of *how to pose the question* in the first place.

In order to deal with the question of the definition of death, even for crass clinical or legal purposes, we need to deal with the ontological question of what a human life is, or as I will put it, what constitutes the life-identity of a human being. And willy-nilly, we have to deal with the epistemological question of how we come to know what such a life-identity is, and therefore, when it ceases to exist.

Before considering how to pose the question, I want to enter several demurrers, or caveats. I reject several distinctions at the outset. First, I reject the distinction between a biological conception of a human death (and *a fortiori*, of a human life) and a socio-cultural or moral conception, on the grounds that the *kind* of biological creatures we are *is* socio-cultural or moral. That is to say, the emergent level of biological existence that concerns us here includes not only upright posture, the opposable thumb, and the cerebral cortex, but how these evolve as conditions of, or even as resultants of a certain form of life activity, which is essentially socio-cultural and essentially moral or normative in character. Second, therefore, I want to reject the distinction between a medical-scientific approach to the definition of death and a social-moral approach (like, for example, the one Kass proposes in his otherwise splendid critique of Morison, some years ago); and this, on the grounds that human medicine and science – down to the biochemistry of the cell – has as its object an essentially social-moral, and also historical being – the human. (Here, I may call on as positivistic a witness as Claude Bernard to testify.)

Third, I want to enter a caveat on the distinction between the natural fact of death, and the socially constituted judgment that death has occurred – not because I think a human death is a matter of determination by social policy or cultural preferences, but because what is at stake in the question of the definition of death is not whether or not death is a natural fact – it surely is – but what sort of natural fact it is, and how we come to determine that this natural fact has occurred. This raises the question of how we are to understand the knowledge of natural facts, or what we take to be true of nature. Finally, I want to enter a stylistic caveat: Whenever I say *death* or *life*, or *human death*, I will always mean *a* human death, *a* human life. The conceptual abstractions refer, ontologically, to individuals only, insofar as we are talking about the facts of life and death; though they refer also to socially constituted and embodied universals, namely, to *judgments* that a death has occurred, or that a life continues. Such judgments are embodied not only in linguistic utterances, or as expressions of beliefs, but also in their institutional embodiments in modes of action or in practices, e.g., burial, mourning, laws concerning property or marriage, etc.

III.

How shall the question be posed? Almost all commentators agree that the question is properly posed as a question about the definition of a human life. The next move is often to assert that this is a "philosophical" question, rather than "merely" a scientific-medical one; and that since medical practitioners are no more specially qualified to make this definition than others, its answer must be arrived at by a different sort of analysis. This analysis will then provide the clinician with guidelines that will lead to the formulation of medical-scientific criteria, which can determine that the conditions of a life-identity are irrevocably lost. But how, and by whom, shall *philosophical* questions be resolved? Why are philosophers any more qualified to make this determination than others? What is the proper locus of judgment on the question of what constitutes a human life? Indeed, is this an expert question?

Debate ensues. Is somatic "life" without self-consciousness or without sensibility a human life? Is a human life defined by personal identity, by psychological processes and connectedness? By social responsiveness? Is personal identity, if that is the definition of life, a matter of individual, personal self-consciousness? I will not enter into this debate here. Rather, I want to characterize it: it is a debate whose resolution will establish some working consensus as to how to go on; it will modify current practice, will lead (and has led) to legislation, to regulations and adopted procedures, to a change in the general social or cultural understanding of what a human life is. In short, it will affect the self-conception of human beings, of their individuality or of their life identity. But if human beings *are*, as I suggested at the outset, the sort of beings who constitute themselves as what they are, in part at least, by what they take themselves to be, then it is *not* the case that human life is simply a given fact, waiting to be properly understood. What human beings are is itself the product of their characteristic life activity; and if a fundamental feature of this life activity is the social and historical self-definition which proceeds through the conscious articulation of this life-identity – in laws, in social and professional practices, in scientific theory, in rituals, in moral codes – then this very identity is an historical artifact which has undergone changes and will continue to do so. This is not to say, however, that the death of an individual is a matter of social determination willy-nilly. But the way in which it is defined, accepted, *declared*, is what articulates it as a human death – i.e., as an object of conscious awareness, about which a judgment is to be made.

The simple fact is that what is at issue is not death as such, but the conditions under which a human being is to be declared dead. This is a trickier point than it may at first seem. We may say, simply, that a person should be declared dead only when he or she *is* dead, and not before. But the emptiness and circularity of such a formulation is obvious. For what is at issue is not *how* we will come to know that, but what it is that we are supposed to know in the first place. And there is no recourse, outside the ring of fire of human judgment, by which we can arrive at the fact in an unmediated way. The conditions of any human judgment – whether in science, in medicine or in morals – are the social processes of decision by which we come to agreement, or continue to disagree. The meta-question is therefore what the appropriate locus of such a decision should be, or what the social space of definition is. The determination that the definition of human life identity, or of human death, should be made scientifically, or medically, or philosophically, or morally; or that it is a matter of the ultimate sanctity of life, or of social cost-allocation, etc., etc., is not a *separate* methodological question of how the question is to be posed, *after* which we may then proceed to answer it, in one or another of these ways. Rather, the choice of one rather than another framework already predisposes a certain determination of what we take life to be, or death to be.

Ontology recapitulates methodology here, as elsewhere: The way the question is posed, and who poses it, already determines in large part what will count as an answer. Suppose the question is posed as one which is to be decided by the clinician. The AMA Judicial Council is clear enough on this. Their report states, "Death shall be determined by the clinical judgment of the physician. In making this judgment, the ethical physician will use all available currently accepted scientific tests" ([1], p. 23). That doesn't tell us that the physician will *define* death, or what a human life is, so that one will know when it's over, but it tells us who is responsible for making the determination, presumably in accordance with some viable and accepted definition. Well, for that we have the various lists of criteria for telling *when* death has occurred, as these develop in medicine, in the law, and in the culture generally.

So far, I have no quarrel with any of this. But who is providing the *definition* for which these criteria are tests? Without this definition, we have what in formal logic is called an *implicit* definition: "Whatever passes these tests is dead." That is, death is *defined* as whatever these criteria test for (as, for example, Intelligence has been defined as what Intelligence tests test). I, of course, am hardly the first to note this peculiarity. But *who* poses the question frames what can possibly count as an answer to it, and interests and understandings and vantage points differ. Recognizing this, and concerned about the zeal and adversarial relation which transplant teams bear to the near-dead, the Committee on Morals and Ethics of the Transplantation Society of the U.S. ("Morals *and* Ethics"! They're not going to be accused of leaving any grave unturned) suggests that "the acceptation of death should be made and declared by at least two physicians whose primary responsibility is care of the potential donor and who are independent of the transplant team." Now this is at least a tacit recognition that the determination *that* death has occurred may be a matter of socially constituted judgment, and that a certain interest (the care of the patient) leads to a better-founded judgment than another (the transplant team's or that of the nephew waiting for the inheritance). But if we take this issue in a larger context, then we may say that the medical interest *in general* is different from the law's interest, which is not to determine death clinically, but to get a clinical determination that will hold up in court, or against plaintiffs; or the hospital administrator's interest, where the social cost of prolonged dying, or of prolonged care, or testing of brain-dead but somatically live patients, affects the economy of health care and the allocation of scarce resources. All this is well known. I'm not introducing it, but I want to use it to make a more general point: that *both* the definition of death and the determination of death are socially bound judgments, which in effect work from tacit definitions of life or of human life-identity; and that what they exhibit is the distinctively human context in which death is a socially constituted fact, and in which a declaration of death is required as a certification that death has occurred.

IV.

Don't get me wrong: I am not arguing that death *doesn't* occur until or unless it is certifiably declared. In the long run, death will occur with or without benefit of certification (or as Lord Keynes said about the "long run", in probability theory: "In the long run, we're all dead"). However, you're not *legally* dead until you are legally *declared* dead; and legal death, like the legal person, in the law is a social fact, or one of the

social facts which the definition of death constitutes. So it would seem
there is death and death: "real", "biological" death, and "social" or
"legal death", and the problem is, presumably, to get them to coincide.
But let me take one more slippery step down the slope: If what
constitutes *human* reality, the *being* of being human, is in large part
social; if we are in part at least what we are recognized as being or what
we take ourselves to be in our culture, or historical setting, then being a
person, being a live human being, is in part determined by what is
socially constituted or defined as a living human being. But then, being
dead, if it is defined in terms of the definitions of such life, is also socially
constituted. Human reality as a social reality, or human ontology as a
social ontology, can't then be reduced to animal biology, and vital signs
must include the signs of distinctively human vitality. That gets us to
consciousness and personal identity, and whole-brain, neocortex, and
all the rest.

The obvious thing to say here has been said often, and here in this
conference as well. I won't repeat it, but just flash the cues: somatic vs.
brain death; somatic death (of the whole system) vs. necrosis (of parts);
whole-brain vs. neocortex; all of which goes to say, being *humanly* alive
is being conscious, or having a prospect of regaining consciousness, or
experience; lacking which, the person is no person and (hard-line)
already dead or (soft-line) not a living *human* being, even if somatically
alive; or one or another permutation or refinement of these. But what I
find fault with here is not the criterion of consciousness, or of personality, or of personal identity, but rather how these have been construed.

My argument, briefly put, is this: The proponents of a neocortical
criterion of human life – i.e., of a definition of human life as requiring
the consciousness which only a viable neocortex can support – tend, in
my view, to take a reductionist, Cartesian view of this consciousness,
and hence of personal identity. *My* personal identity, *my* consciousness
is certainly my own, and no one else's. The principle of individuation
here is certainly *self*-consciousness, embodied uniquely. But my self-
consciousness is not something I came equipped with. It is a product of
birth and maturation in a social environment in which it is dependent on
recognition by another self-consciousness like my own.

Self-consciousness and personal identity have as their conditions
social interaction, social recognition. Now I am *not* heading where it
looks like I'm heading. I am *not* going to argue that, therefore, the
condition for my being declared a human being, a person, is my social

responsiveness, or my ability to communicate with or recognize others. I am rather arguing that *my being conscious* in *this* sense is constituted in large part by *my being taken to be a person by others*. Even where I fail to maintain my self-identity – as in pathological mental states – I remain the person I am *if* I am alive and am recognized by others as that person. As a social being, I am not simply a Cartesian *I*, but (as for Hegel) *the I that is We, and the We that is I*. Therefore, though I may never regain my conscious awareness of others, if I am somatically alive, and they recognize me in that state as being the person I have been identified with continuously to that point, then, I would argue, I continue to *be* that person in a pathological (and irreversibly pathological) mental state, since my personality continues to be constituted by my continued recognition as that person by others.

Now *whether* I am so recognized or not depends on how I am taken to be; and for myself, I would argue that I would want others to continue to take me to be that person, as long as the breath of life is in me, but no longer, if I cannot persist alive without heroic measures. I don't take this to be a *moral* preference, but an ontological fact about the *being* of being human, i.e., the sociality of human personality. The rights I have to this continued recognition derive from my *having been* that person, so I don't think this applies simply to the fact of my being a viable, living organism *tout court*. (Thus, I don't think this holds, at the other end, so to speak, for foetuses, because they are not persons, but that is a difficult and separate argument.) I am, in that state, a residual personality, so to speak; and it seems to me that it would be odder to say that my still living body is no longer continuous with the living body it was, in terms of its identity, than to say that my bodily identity – the body that bears all the marks of my experienced life, and still lives – is a corpse.

Would this be an extraordinarily difficult and painful burden for those who continue to recognize me as a person? Certainly. Should I be permitted to die, short of heroic measures? Yes. Am I no longer a human being? No. My humanity persists as a social, if residual, fact. Should my death be rationally wished for by others? That, I would say, is up to them. Rationality is too fragile a reed, and too ill-defined a human trait, to have it determined in such a case for any of the living by any others than themselves. Social cost-accounting is not, I think, rationality in such a case.

What I have proposed here is an anti-Cartesian notion of consciousness. I have not proposed the perverse claim that I, in that state, would still in some weird sense be conscious, or an experiencing or responding subject; but that the consciousness *of me* as in that state yet living, constitutes my continuing self-identity as a living being, as a human life-identity without self-consciousness. Even then, as somatically alive and neo-cortically dead, I am not mere animal organism, but human being. My history persists in my presence.

The City University of New York
Baruch College and The Graduate Center
New York City

BIBLIOGRAPHY

1. American Medical Association Judicial Council: 1977, *Opinions and Reports*, AMA Press, Chicago, Ill.

CHARLES E. SCOTT

THE MANY TIMES OF DEATH

Professor Wartofsky is right to emphasize that we are social and historical beings who lose our being if we are abstracted from our specific
societies and histories. He is also right when he says that criteria by
which we determine when we are dead are significant in the processes by
which we form ourselves and give ourselves identity. If, for example, we
abstract our deaths from our social and individual history, we build into
our society a severance of our lives and deaths from our own specificity
and concreteness. We make a schizzy kind of alienation for ourselves,
sew it into the fabric of our lives, and make certain that we will not be
able to feel attuned to our own lives in the ways in which we live and die
them.

Consider the Commission *Report*'s own social significance. Its function is to provide a definition for death that is to be followed in our
medical institutions and in our courts of law. It is designed to make us
feel right about terminating life supports when the whole brain stops
functioning. It responds to popular and political concern for statutory
uniformity regarding death-determination. It seeks out a low common
denominator of conviction having to do with the determination of death
in order to make its recommendations politically feasible. It takes
account of – indeed we could say that the *Report* is conceived in – the
difficult, competitive interplay of governmental and medical concerns
over authority and professional self-determination. And it uses a
method of determining when death occurs that abstracts a person or a
body from the person's or body's history, culture, and life-world. The
Report is an exceptionally fine example of a society-forming document
that ignores the social identity of its subject matter, dying and dead
people.

This kind of abstraction from the ways in which deaths are lived in
specific settings will cause cynicism and confusion. It will advance an
already existing distance between our knowledge and experience of
ourselves and our cultural self-understanding. The consequences of such
a rift minimally are institutions that distance us from ourselves as they
serve us. These institutions generate ideals that mislead or hinder what

229

Richard M. Zaner (ed.), Death: Beyond Whole-Brain Criteria, pp. 229–231.
© 1988 *by Kluwer Academic Publishers.*

they are designed to lead and nurture. When people are distanced from their own traditions and ways of living by a society's practices and procedures regarding death and dying, something like a cultural schizophrenia results: the better those institutionalized procedures function, the more alienated individuals are from themselves.

Professor Wartofsky's paper suggests a significant option to the Commission *Report*'s,proposal. The *Report* is single-minded in establishing a uniform termination of death. Its assumption is that *one* thing dies when a person dies and that one kind of determination can establish when all human deaths occur. It assumes that death is like a single identity – a peculiar assumption when you think about it. Professor Wartofsky uses his social-cultural understanding of human being to show that the dying individual, this person, cannot be separated from her specific relationships, his traditions of life, her being recognized and known in particular ways. If such a separation occurs, the life and being of the person is obscured. Wartofsky's metaphysical claim is that human beings are essentially social-historical beings. That means that appropriate and non-alienating criteria for determining death must address not merely the "animal death" of people, i.e., the cessation of biological functions. The criteria must address us as we are, as beings who are human because of specific social relations and living histories in the form of practices, traditional duties, and complex patterns of feelings, attitudes, beliefs, and ideas. Human beings are not defined by internal states of consciousness, but by an extensive life-world, by the judgments regarding them, and by the social fabric that defines their cultural space and time.

This social-historical understanding of human being means not only that the "fact" of human death is constituted by social judgments. It means further that no one way of dying is definitive of human beings. Not only is biological or "animal" death only one dimension of human passing away. In the essentially human dimension of dying, many different definitions of death are legitimately possible. Different social-historical patterns – different traditions and subtraditions – mean that a pluralization of the time and definition of death is necessary if we are to treat human death appropriately.

Professor Wartofsky is clear about his own preferences. He says that if his friends hold him with respect they will not use heroic measures to keep his body alive even when he is neocortically functional. By implication I can say, consistent with Professor Wartofsky's position, that if I and my friends want my body sustained by all possible measures, that is up to us and the values and practices which define "us". There are other legitimate, different options we might think of. The point is that one's history, one's regard by others, and one's persistent self-regard all constitute one's human life. I understand this claim to mean that different groupings of individuals will constitute different times and definitions appropriate for death. "My history persists in my presences," says Professor Wartofsky. Human life persists in the individual's social fabric (i.e., its presence) and no where else. Given the varieties of social fabrics, human lives have multiple definitions, and deaths will be quite different as to when and how they occur. What is human death in this history will not be human death in that history.

The belief that death is reasonably to be assessed by a unified definition is one mistake to avoid if we wish to understand our diverse selves in and through our dying practices. Our best point of departure is to be found in our diversity, our cultural selves, our living histories. Our greatest danger is not murder of bodies, but radical mistreatment of human beings. One of those dangers can be seen in the effort to provide criteria for life and death by reference only to biological functions conceived with an emphasis on singularity and unity. Professor Wartofsky began by vowing not to deconstruct death. He did succeed nonetheless in suggesting that practices and knowledges centered in a unified definition of whole-brain death must be deconstructed by a way of thinking that respects the essential diversity of human lives and deaths. The next step is to show how medical and legal knowledge related to death and dying have ignored this diversity – to deconstruct these regions of professional knowledge – and to find out how to make our diversities part of our legal and medical understanding of who we are as we die.

Vanderbilt University
Nashville, Tennessee

JOHN LACHS

THE ELEMENT OF CHOICE IN CRITERIA OF DEATH

CAN WE *FIND* THE POINT OF DEATH?

Death is easier to undergo than to understand. It comes unbidden or we can attain it with minimal effort. Yet the willfulness of human nature makes it difficult for us to settle for the easy; it is understanding of it we want, not the experience.

Thought begins with the silent assumption that words reveal the world; the unitary noun "death" must, therefore, be the name of a single phenomenon. Since we can readily distinguish living persons from decomposing bodies, it is natural to suppose that death is something interposed between these two, a happening that rends the smile of life and reduces us to a heap of cooling cells. But if "death" denotes such an event, it must occur at some precise moment between the time we still thrive and the hour our remains are interred. Lives terminate at some point, we all seem to agree; there must, therefore, be a simple and correct answer at any given time to the question of whether a person is then alive. If he had been born and has not died, he is still with us; if he died, we should be able to determine how long he has been gone. Death, then, appears to be a point-event that separates us from the dark. It is a single, simple, irreversible happening that has temporal location but no temporal dimension: it occurs at a time, yet it takes up no time in occurring. It is an instantaneous affair.

This is not only where thought begins, it is also where it ends for many earnest people. Physicians tend to conceptualize death in this way and find support for their view (when they think confirmation is needed) among certain philosophers. Even the President's Commission endorses this idea in drawing a distinction between the temporal process of dying and death as an instantaneous event. It quotes with approval the view of Bernat, Culver, and Gert that "death should be viewed not as a process but as the event that separates the process of dying from the process of disintegration" ([1], p. 77). I readily acknowledge the attractiveness of this approach. We are, after all, either dead or alive, and which of the two appears to be a matter of empirical fact. If so, it should be possible

233

Richard M. Zaner (ed.), Death: Beyond Whole-Brain Criteria, pp. 233–251.
© *1988 by Kluwer Academic Publishers.*

to articulate the correct criteria of death, which the President's Commission promptly undertakes to do.

We should note, however, that such simple and attractive views carry a danger. They seem so right, they tend to paralyze the mind. They stunt critical inquiry and thereby make it difficult to uncover their hidden commitments. Since even the simplest views are based on assumptions, the result is inadequate self-understanding on the part of those committed to them and general inability to give them a just appraisal.

The theory that death is simply an event in nature has its own central assumptions. For if it is a purely biological phenomenon, our proper cognitive relation to it is that of discovery. Its status then resembles that of America before Columbus' arrival: it is there ready for the light of knowledge to fall on it, ready to be detected and explored. To be sure, our continent was not *called* America before the Europeans arrived. But discoverers add only the name, they do not create the lands. So it seems to be with death also: it is a natural phenomenon we discover and then name. We contribute nothing to its existence; on the contrary, it is its prior reality that makes our inquiry possible and useful. Our initial, natural view commits us in this way to a realist ontology concerning death.

THREE SORTS OF CASES

To see how much of an unjustified presumption is involved here, we need to distinguish three different sorts of cases. There are some instances, such as that of the Pacific before Balboa beheld it, in which objects enjoy existence independently of any cognitive or conative human act. In a radically different set of cases, existence is totally the outcome of our activities. The meeting of a Board of Directors, for example, is an occurrence altogether dependent on acts and practices. For no one is a director except in a complex social context and no meeting of those who are directors is a meeting of the Board unless its scheduling and convening follow established rules.

There are also intermediate cases in which human choice and action build on preexisting things or conditions to create a novel object. No person is fat apart from human judgment, yet corpulent people are not created by our perception or our whim. There is an underlying physical reality: the distribution of weight among the population. This continuum, ranging from those who weigh less than 100 pounds to individuals who tip the scales at over 500, then receives a human contribution which breaks it into the loose categories of the thin, the normal, and the fat. Such categorization, based on taste, tradition, and social purposes, is by no means trivial. It establishes new objects, namely, groups of people judged to be deserving of differential treatment medically, socially, and in our personal relations.

In the human world, there are a very large number of cases of this third variety. The law and our social practices frequently call on us to draw relatively sharp lines to separate phases of an otherwise continuous process or to stress the differences among remarkably similar conditions. Such differentiations represent human contributions to physical facts: they express our choices and embody our values. For the man who wants a fine lawn, it is important to kill noxious growths. But, in its physical constitution, nothing is a weed. Weeds are physico-social objects, viz. broad-leafed plants that grow undesirably in the grass. The combination of independent existents and selectivity expressing human interests and purposes is evident here, as it is when we call someone smart or beautiful, or when we set the time at which a young person can begin to drive or fix the end of the Middle Ages.

In which of the three types of cases does death belong? Although some people, Christian Scientists perhaps, view it as a human creation or illusion, it is unlikely that death lacks objective basis. Does this mean, however, that it is an exclusively physical event which "separates the process of dying from the process of disintegration"? Although this is the currently favored position, I do not think that it is right. Persons of scientific sophistication should have long had their suspicion aroused by the notion of death as an instantaneous point-event, for such presumed occurrences resemble humanly contrived termini much more closely than they do the unbroken temporal processes we find in nature. The function of determining the *exact moment* of death yields another bit of evidence against the purely physical view. Our interest in this is largely legal, not scientific: we want to know in order to establish criminal liability or, as in the case of multiple deaths in a family due to a single accident, to clarify the pattern of inheritance.

DEATH A PHYSICO-SOCIAL REALITY

The role this notion of the death-event plays in our common practices provides added confirmation that the object it identifies is not a purely physical reality, but a physico-social one. Our primary purpose in applying the concept of death to people is to indicate a very important point of change in our relations to them. So long as people are alive, certain activities are obligatory while others are forbidden with respect to them. If we are members of their family or of the team that cares for them, we must, for example, attempt to communicate with them, to ascertain their wishes and to aid them when appropriate. On the other hand, we are not permitted to open their safety deposit boxes and read their wills, to perform autopsies on them or to cart them to the cemetery for burial. The relations and the activities that embody them change in a radical way once death has taken place. It then becomes acceptable for us to treat our patients and loved ones in the ways our society deems appropriate to the dead.

No one would want to deny, of course, that there are important differences between those alive and the dead. These divergences constitute the objective foundation for distinguishing the two. But how much difference there must be and which traits or activities of the living must change in order for us to say that they have died are matters of judgment. Since the concept of death functions as a trigger or an indicator to alter our activities with respect to those to whom it is applied, it must involve a social standard of decision about when such change of behavior is proper.

In developing criteria of death, then, we must do two things instead of the one usually supposed. On the one hand, we must attend to the facts of organic decline; this is an empirical inquiry about independently existing biological phenomena. On the other hand, we must also formulate a socially acceptable decision about precisely where in the continuum of decay to draw the line between the living and the dead. This is essentially a normative activity in which we bring tradition, religious and moral values, and social utilities to bear on important human relations.

VALUE ELEMENTS IN THE CHOICE OF DEATH CRITERIA

The current line of thought sheds new light on the conflict between the whole-brain and the neocortical criteria for declaring a person dead.

Advocates of each position tend to view the controversy as factual, namely, as a disagreement about when or under what circumstances death *in fact* occurs. It should be obvious by now that there is no factual solution to this debate and that both sides, when they insist on one, betray inadequate self-understanding. There can be no objective solution because there is no empirical disagreement; neither side offers an alternative biology. The real opposition, though misleadingly couched in factual terms, is conceptual and that conceptual conflict is grounded in incompatible value-commitments. Advocates of the neocortical view tend to think of humans in terms of the category of persons, and thus regard death as the cessation of such higher activities as feeling, consciousness, self-consciousness, and reflection. Proponents of the whole-brain thesis, by contrast, adopt the ancient and perhaps more inarticulate idea that humans are intimately connected with or inseparable from all of their integrated bodily acts.

This conceptual disagreement is but the cognitive expression of divergent value commitments. Champions of the neocortical view prize consciousness and the activities it makes possible above all else. The vehemence of their affirmation that the life that lacks awareness (and related acts) is worthless sometimes takes the form of refusing to call biological existence without neocortical function "life" at all. Their reasoning proceeds not from the objective recognition that cessation of higher brain functions is tantamount to death. Instead, though often only tacitly, they move in the opposite direction: initial commitment to the all-importance of certain activities makes them choose destruction of the organs supporting those activities as the indicator of death.

Advocates of the whole-brain criterion think in a similar fashion. But their devotion, no doubt traditional and religious in derivation, is to integrated bodily acts. Factual considerations come in only when they try to establish the biological foundation of those acts. Cessation of the integrative work of the entire brain, then, is not the sign of when death *in fact* takes place; it is where we must place the point of death if we have an antecedent commitment to the importance of integrated bodily function.

We cannot, therefore, adjudicate the conflict of criteria in anything less than frankly normative terms. The question to ask is not when death really occurs, but why we should draw the line between life and death at the higher cognitive rather than at the lower integrative activities. Recast in these terms, we can develop a better understanding of the differences between the criteria, and hope eventually to resolve their conflict in a socially acceptable way.

BIOLOGICAL FACTS, SOCIAL DECISION

The essence of my claim is that formulating a criterion of death involves, in addition to factual considerations, a social decision. Whether the heart beats is an empirical matter susceptible of factual determination. Whether when the heart stops beating death has occurred involves the application of a social standard. This standard is not written into the nature of things and, accordingly, cannot be read off from the facts of biology. It is established by taking biological realities into account and then making a more or less conscious communal choice. The choice is not dictated by what we know of the operations of the body; religious, social, moral, and technological considerations also play a part. That, for example, we insist on the *irreversibility* of the cessation of the heart, lung, or brain function is not a matter of recognizing what death is; it is an expression of our interventionist values. What it means is that we refuse to view anyone as dead so long as there is something we can do to sustain or restart their biological machinery.

In other cultures, the normative and religious milieux make cardiac resuscitation unthinkable; when the gods stop life, such people might believe, it is sacrilege for humans to interfere. If someone responds by claiming that our practice is superior, he yields the point at issue by moving to the normative plane. For then he agrees that criteria of death are far from being purely factual: if irreversibility, for example, is not a part of them, then it ought to be.

OTHER SOCIETIES

If I am correct that locating death along the continuum of organic decline involves a social choice, it is likely that different societies place it at different points. And that is precisely what both history and anthropology show. People in some societies have, not unreasonably, decided that no one is dead until the next sunup after spontaneous movement ceases. In other cultures, expressing different values and divergent natural constraints, death has been pegged at the point where a person gets sufficiently weak to consume more social goods than he can produce. Primitive Eskimos might have had something like this in mind when they invited infirm old people to lessen the community's burden by walking into the blizzard or the night. Cruel and unacceptable as this might seem to a society in need of consumers, it was clearly reasonable and legitimate under then prevailing circumstances. I note that even there death had a physical basis, but it was supposed to occur relatively early in the process of decline and disintegration. The act of walking out into the snow was not suicide but proper acknowledgment of the fact that, in the eyes of the community, one was already dead. Along with whatever weaknesses it displays, this view has an intriguing and beautiful feature. In industrial society, our last act is an involuntary movement of the heart, lung, or brain. Eskimo culture permitted many the privilege of a last act that was not only voluntary, but virtuous and noble.

My point is not to urge the superiority of Eskimo ways. I used their example to underscore the fact that the point of death can be situated at a number of places along the continuum of organic decline. Where we put it is a matter of choice determined by the beliefs, values, and circumstances of the community. It is this choice that the naively realistic view of death obscures. Even the most fervent belief that death is something the criterion of whose occurrence can be discovered or read off from the facts cannot eliminate the element of social decision involved in formulating it. The naiveté of the realistic view, well exemplified in the report of the President's Commission, succeeds only in keeping the choice unconscious and hence uncriticized. The thought that we can *discover* the point of death conceals from us what we really do and thereby makes an intelligent examination of our activity impossible.

DEATH: A BIOLOGICALLY BASED SOCIAL STATUS

My argument shows that we have reason to think of death not as a simple organic condition but as a biologically based social status. Since the status is social in nature, there are public and community-wide standards for determining it. This is the truth behind the claim that the decision of when one is dead cannot be left to personal whim. The choice is not, therefore, that of the individual physician or of the patient. It is a choice the community must make, or has made, as displayed in its rules and practices. The physician merely applies this customary or statutory standard.

It is important to remember that the presence of human decision does not remove generality, only the comfort of supposing that our ideas are replicas of reality and hence our beliefs accord cozily with the nature of things. The choice, moreover, is obviously less than absolute: no society is free to set as the sole criterion of death a sneeze or a strong itch under the left armpit. The line of death must be drawn at some significant point on the continuum of personal decline. But such disintegration involves the change or cessation of a multiplicity of activities; societies have considerable leeway in pinpointing death at the stage where functions of particular interest to them become impossible to perform.

ACTIVITIES AND COSTS

If the line between life and death is drawn on the basis of a social decision, what are the factors that influence this choice? There are two considerations that appear to be always present. The first is, as I have just indicated, the identification of important activities which, when one can perform them no more, is a sufficient reason for thinking one dead. Declarations of death are, therefore, at once (though tacitly) statements about what the persons involved are no longer able to do. The second consideration is the assessment of our responsibilities to declining people and of the cost of carrying them out. We have many duties to people who are alive; declaring them dead instantly eliminates a host of these obligations, imposes only a few new ones, and makes hitherto unacceptable acts appropriate.

Lest I sound cynical in mentioning the cost of locating death at one place or another in the biological continuum, let us remember that societies that are hard pressed have always set the point of death on a cost-benefit basis at the early stages of non-functionality. In many contexts, elderly Eskimos could still function reasonably well. But the cost of supporting them through debility and disease was thought too high for the meager resources of their community. When Europe suffered from the Plague, individuals with the first signs of the disease were treated in ways appropriate to the dead: the health costs of providing treatment or even the comfort of companionship were judged intolerable. The long-term maintenance of numbers of humanly non-functioning biological wrecks is a luxury open only to rich and stable societies. I hesitate to infer from this that none but the very wealthy can do what is morally right.

The growth of technology and the rapid increase in the number of old people in our own society may make cost considerations paramount again. If the number of individuals with severe neocortical and other organic damage but functioning brainstem were to increase dramatically, another President's Commission may well find the cessation of higher brain activity an adequate criterion of death. As an absolute minimum, escalating health-care costs are likely to cause a reconsideration of the current requirement of the irreversibility of the cessation of key organic functions. Our actual practice of refusing resuscitation to significant numbers of terminal patients, clearly at odds with the Presidential Commission's criterion, already shows recognition of the deeper wisdom that although some heart stoppages may well be reversible, from a human standpoint they are not worth reversing.

It should be clear from these comments that the cost of what we would have to do for certain groups of people if they were judged to be alive is a factor, and sometimes the decisive factor, in formulating a criterion of death. But in many if not most cases the primary consideration in the social decision underlying the definition of death is what individuals at different stages of organic disintegration can do. The issue is in part one of technology: with the aid of suitable devices, we can now do more over a longer period of time than ever before. But for the rest, it comes to a question of our beliefs, values, and needs. For it is these that shape our view of the hierarchy of human activities.

What are the acts and achievements, the absence of which has been supposed to render one dead? Even if we leave out the personal and idiosyncratic failures of function for which individuals have rendered themselves dead, we find a bewildering variety of activities. I have already mentioned relative economic unproductivity and the early signs (with minimal loss of function) of contagious disease. To this we can add loss of rational control over one's life, loss of strength sufficient to prevail in combat, permanent loss of consciousness, loss of pulmonary function, loss of cardiac activity, and loss of the integrated operations of the entire brain. At the spiritual end we find irreparable damage to the proper work of the soul or to one's relations with God. At the other extreme, death is supposed to occur only when all biological activity, down to the cellular level, has ceased or when putrefaction sets in. There is hardly any significant human activity that some society or sub-culture has not thought crucial to life and whose loss has not been supposed to signal death.

ISN'T DEATH A BIOLOGICAL FACT?

An objection is likely to be urged, perhaps quite impatiently, at this point. Those who insist on the pure factuality of medical science may indicate displeasure with the way I seem to mix the social and the biological, what to them appear as the optional and the compulsory. This tends to blur the distinction between ostracism, traumatic or destructive community practices, metaphorical senses of "death", self-sacrifice and suicide on the one hand, and the stubborn reality of the cessation of biological function on the other. Modern medicine takes little interest, I may rightly be reminded, in odd social rituals. Its gaze is fixed on the natural process that underlies human interaction and whose end destroys its possibility. Whatever a community does to or requires of its members, their actual death does not occur until their bodies give out. The definition of death should, therefore, reflect this primacy and ultimacy of the physical.

This is a serious objection and, although much of my discussion so far has been devoted to showing the dubiety of its assumptions, it deserves a straightforward response. I begin by happily granting the importance of biological process. My concern is only with our current supposition of its all-importance. Life is a very large collection of activities; the cessation of some, many, or most of these constitutes death. I want to stress that not *all* the operations of life cease at any point where death is supposed to occur: some residual activities and relations go with us to the grave. There is, therefore, significant selectivity in deciding which activities are central for life and which are peripheral only. The President's Commission, for example, maintains that brain-integration of biological function is essential; its absence is a sure sign, or is even the very nature, of death. By contrast, apparently on the theory that cellular activity is necessary but not sufficient for organized life, its continued presence is declared irrelevant to the death of the organism ([1], p. 28). Those who favor cessation of neocortical function as the criterion of death, on the other hand, insist on the centrality of the cognitive and conative activities associated with consciousness. To them, continued organic process, even if system-wide, is insufficient to keep persons alive.

THE CHOICE OF WHICH LIFE-ACTIVITIES ARE CENTRAL

How are the "essential" operations selected from among all those that compose human existence? All choice presupposes interests or purposes or values. The focus of the physician is the biological substratum of life, its roots and not its flower. Understandably, therefore, he gives preference to causally central, capacitating processes. Are these in fact the crucial operations of life? Almost certainly, if we identify life with organic existence and thus adopt the biological perspective of medicine. But if we seek the fruit of life and not its sustaining causes, the view of medicine may well appear reductivist and inadequate. The activities selected as central to life will then be thought, emotion, and the other higher functions; their termination will constitute death just as really as cardiopulmonary failure does for the physician.

We can see, then, that the selectivity necessary for identifying central life-activities inevitably carries us past the factually empirical. It introduces a set of interests and a system of correlated values. The biological perspective common to physicians is, moreover, itself a social creation with its own underlying purposes. It is by and large the collective possession of an entire, important subculture in our society. And this society permits, encourages, even fosters the predominance of the biological perspective in the medical subculture because of the immense health benefits it provides.

None of this presents a problem if it is properly understood. But the precise factuality of science and vast success of medicine tend to make us forget about the interests and the selectivity that underlie them. Expertise is easily converted into truth: insensibly we drift into thinking that biology and biochemistry and medicine provide an accurate, literal, and complete account of human reality. Medicine becomes, in this way, the final arbiter of life and death: we look to it for precise information, for solutions to our problems, even for reliable standards of when life, and not merely its biological support, ends.

STANDARDS OF HEALTH AND DEATH SOCIALLY ESTABLISHED

When we convert the results of this important but conditioned and limited enterprise into unconditional truth, we lose sight of the vital role social decisions play in establishing it and shaping its structure, and in setting the standards of health and death. This inattention is the source of the false idea that death is a purely natural terminus which involves no human activity other than the attempts of the health care team to retard or to reverse it.

In fact, we cannot understand death and cannot even hope to establish rational criteria for it without putting it in its proper historical and social context. If we do that, we see at once the omnipresence of human activity, the way in which our choices and values surround, frame, structure, channel, establish, accelerate, and in some cases eliminate natural processes. The best way to remind ourselves of this centrality of human purposes is to call attention to the diversity of social practices. The absence of this cross-cultural or anthropological perspective renders much contemporary discussion of the criteria of death one-dimensional and the deliberations of the President's Commission sterile and unsatisfying.

My response to those, therefore, who charge that in discussing death I mix social and biological considerations is warm congratulations. They have detected my strategy; I hope that at this point they can even see my arguments for it. If they do, they may come to agree with me that we are unlikely to make progress in this field without due attention to the social and normative context, that is, to the deliberate choices we must make and the reasons for them.

HOW TO CHOOSE A DEATH CRITERION?

How does paying heed to choice and value help us in assessing the relative merits of the whole-brain and neocortical views of death? First, it protects us from the misleading suppositions, fatal for inquiry, that the primary issue between these competing theories is either factual or conceptual. It is, of course, absolutely essential to get the relevant facts exactly right. And conceptual clarity and precision are equally important. But these are initial requirements only; without agreement, which is hammered out of the traditions and beliefs of the community and which takes due account of social utilities. they are inadequate to establish the line between life and death.

In our society, the agreement is achieved within the confines of the political system, taken in its broadest and most inclusive sense. This system can operate only if public discourse is free and enlightened. Such discussion of community policy, in turn, presupposes accurate information and an educated citizenry. The role of physicians is not to inform us of when death *in fact* takes place, but to provide reliable knowledge of the stages of organic disintegration. And the job of philosophers is not so much to resolve thorny ontological problems about the personhood of human beings, as to focus public discussion and political decision-making by the critical examination of our beliefs and of the consequences of the values we embrace.

The *Report* of the President's Commission was political in just this sense: it offered itself as a voice in the social dialogue, presenting a criterion of death that might be generally acceptable. But, lacking a clear understanding of the central role of values in any such proposal, it attempted, erroneously, to pass off its results as supportable by scientific facts. Such a naively realistic approach may be thought to be rhetorically effective: who, after all, would want to contradict what physicians say about the objective realities that fall within their domain? Yet the conclusions of the Commission have succeeded neither in enlightening public discourse nor in stilling the controversy surrounding the conditions under which a person should be viewed as dead. Lasting agreement will not be achieved until we develop a reflective and public consensus about the relative costs and benefits of rival positions.

WHOLE-BRAIN OR NEOCORTICAL VIEW?

This is obviously not the occasion to work out the details of such an assessment. I will sketch only its outlines, calling attention to the sorts of considerations we must take into account. I have already indicated that these fall into three broad categories.

1. *The Weight of Tradition*

First, any attempt to establish the line of death must reckon with our habits, traditions, and established practices. These appear to be a mixed bag, with some favoring the more conservative whole-brain criterion, while others are readily compatible with the neocortical view. We have, for example, the deepest devotion to the human body, caring for it and treating it with respect well past the point where such regard can do the person inhabiting it any good. We also have habits of optimistic intervention, which make it difficult for us to admit that anyone is past hope and nearly impossible to feel released of the obligation to continue our attempts to benefit him.

Such practices support the view that death cannot be thought to occur until centrally controlled organic activity ceases, or perhaps until even later. But other established forms of behavior point to the permanent loss of the higher conscious functions as the proper dividing line between the living and the dead. Our spontaneous attitudes to people do, after all, undergo radical change when we find them unresponsive in a deep coma. And those closest to patients so afflicted, the members of their family and of their health care team, tend to deal with them in a way that is clear-eyed and unsentimental. This is amply displayed in the widespread practice of denying resuscitation not only to comatose individuals, but also to those in advanced stages of senility and to terminal patients in drug-induced stupor.

2. *The Role of Established Beliefs*

Our shared beliefs show the same ambiguity as our practices. On the one hand, we have such a strong commitment to the value of individuals that the possibility of false positives – the survival, reversal, or misdiagnosis of the loss of higher conscious functions – makes us suspicious of the neocortical view. Although nineteenth-century fears of burying people who seem dead but are somehow alive are no longer with us, it is still difficult for us to believe that individuals who appear physically intact and may even breathe on their own could, nevertheless, be irremediably dead. On the other hand, we also believe that people who have permanently lost consciousness have undergone a profound change and that respectful treatment of their bodies is largely the expression of residual respect for who they had been. When such treatment becomes protracted and burdensome, the people directly involved with it — who tend to understand the situation best — generally agree that though the body may linger, the person is no longer there.

Our beliefs in these matters were formed under the influence of our religious heritage. But the Judeo-Christian tradition is itself of two minds about the significance of the body for true life. It invites us to consider our physical nature as but the earthly shroud we should be happy to shed or as a prison we must be eager to escape. At the first reliable sign, therefore, that the human (not the vegetative) soul has fled, we are entitled to dispose of the body as we would of any other piece of unneeded matter. Yet, another powerful element in the tradition insists that the connection between person and body is far more intimate. If, in this world at least, the soul is inseparable from the body, respect for the person is respect for his or her earthly form. It is, moreover, not for us to decide when the physical tenure of persons comes to an end: we must support them without fail until God's will is made manifest by the total ruin of their bodies.

3. *Social Utility*

The line of death we draw should be conservative in the sense that it agrees, to the greatest extent possible, with our practices and beliefs. My discussion so far makes it clear that neither of these clearly favors the neocortical over the whole-brain criterion. The deliverance of the third central consideration– social utility –is, however, almost univocal. In every respect but two, drawing the line at the total cessation of the integrative activity of the brain entails vastly more cost than its rival. Of the two, one is purely economic, namely, the employment by the hospital industry of a significant number of people for the purpose of caring for permanently comatose individuals. Given the availability of more productive, alternative employment, this is not a compelling consideration. The other cost of a neocortical criterion is the possibility of error it presents. It is, after all, conceivable that people who appear to have lost all higher conscious functions may, at some later time, astoundingly regain them. But additional vigilance readily reduces or eliminates this cost: the criterion must be so phrased and applied that the recovery of anyone about to be declared dead is only a logical, not a clinical possibility.

For the rest, the neocortical view is vastly preferable. The suffering of patients who might be capable of feeling pain but not of taking or asking for relief, the protracted torture of their families, the sensed impotence of the health care team, and the unproductive use of social resources all point to the unwisdom of the whole-brain view. As any observer of the current scene knows, social choice between the two criteria is not easy. But if they are our major alternatives, rational assessment favors destruction of the neocortex as the better place to draw the line of death.

THE INDIVIDUAL-CENTERED MULTIPLE CRITERION VIEW

Are these views, however, the only major contenders? There is another one that has, for the most part, escaped the notice of philosophers. That it enjoys broad support is demonstrated by the spread of legislation permitting people to write living wills. Because we tend not to think of the criterion of death as chosen, we have failed to recognize the remarkable fact that such laws empower each individual to adopt his or her own criterion. The multiplicity of criteria from which persons can freely choose is necessary for the good death and particularly appropriate to the political system we have adopted. The matter is complex, so here I shall only sketch the nature of and the arguments for this individual-centered multiple criterion view.

If our exertions are mere motion without consciousness, there is no such thing as the good life. Similarly, if departing this world is a clinical point-event, there can be no good death. Although there are clear asymmetries between the good life and the good death, there are certain important similarities due, in large measure, to the presence in each of what makes for the good anywhere. At any rate, the good life and the good death form an integral whole. This is well expressed in Aristotle's view that a timely and fitting exit is necessary for the completion of a rich and satisfying existence. It is difficult to think of happiness, in fact, as constituted by less than just such a structured and meaningful life. The sensible conclusion to which this leads is that death is a part of life (even though it may well be the last part), and that the good death is an indispensable element in the happy life.

THE GOOD DEATH

Innocent as this view may sound, it has important political consequences. For, combined with our inalienable right to the pursuit of happiness, it implies that we have a similar right to seeking the good death. Mindless legislation banning suicide violates this right. It is astounding that the President's Commission, though entrusted to deal with *ethical* problems, shows no awareness of the distinction between death and the good death, of the relation of the good death to the good life, and consequently of the right of individuals to self-determination in the matter of when, how, and under what circumstances they shall expire.

For those who do not favor the rhetoric of rights, the same point can be made in different language. In our country it is established public policy to promote such central values as liberty. Freedom cannot exist, of course, without limit. But the spirit (though not always the letter) of our laws, well supported by fundamental moral principles, is to restrict the liberty of each only where it infringes on the liberty or welfare of others, and to leave it unabridged in matters relating to the individual alone. Since death is paradigmatically a private matter, it follows that its determination should be left open to individual control to the greatest extent possible.

On a very general analysis, the good life has three basic conditions or ingredients. We must have desires and purposes, we must be fortunate enough to live in circumstances where they can be satisfied, and we must have the capacity and energy actually to achieve them. Without the convergence of will, luck, and power, the good life is impossible; even Stoic control and Eastern resignation presuppose certain aims and, if nothing else, the luck of having the internal strength to crush worldly desires. The lack of power yields envy and bitterness; the absence of will makes existence meaningless; and without luck life remains a frustrated quest.

The good death has the same conditions as the good life. First and foremost, we must be fortunate enough not to pass away before our time. We must avoid extravagant cravings, such as that for endless physical life, and frame sensible desires about the time and manner of our demise. Finally, we must have the power to achieve what we want either by direct action or by causing others to respect our will.

The coincidence of will and power makes for effective autonomy. The role of the state is to respect and to safeguard this autonomy in the choice of death no less than in the decisions of daily life. To afford people control over their own death amounts to giving them power to determine how they are to be treated. Autonomy in the choice of conditions under which we no longer wish to live, therefore, requires our legal and social empowerment to decide when members of our health care team are to cease their labors and begin to treat us in ways appropriate to the dead.

CHOICE OF ONE'S OWN DEATH CRITERION

If my earlier argument that death is a biologically based social status is correct, to say that someone is dead is partly to affirm that he has reached a certain stage of organic decline and partly (and most importantly) to indicate a change in our relations to him. As a signal, the declaration is meant to shape our expectations and to revise our obligations. Such statements must not be made lightly or without warrant. What justifies them is an assessment of which important activities the individual in question can no longer perform. Who is to determine what these valued activities are? The decision is best left to the person involved: only she can judge what is of paramount importance in her life and when her existence is no longer worth its cost. In a pluralistic democracy we can acknowledge this sovereignty of the individual over her life and leave the timing and manner of death to personal choice. Instruction by a competent individual of his fellows about the circumstances under which he no longer wishes to be treated as a living person does not require an instant funeral. There are stages in our dealings with the dead; the initial declaration, though it precipitates complex social and legal changes, may demand only the cessation of treatment or, more controversially, active steps to bring a recalcitrant heart in line with its owner's will.

There must, of course, be limits to such self-determination. The decision about the timing of one's death, though not irrevocable, must be serious and sincere. And there has to be an appropriate biological basis for the choice: those suffering from sinus colds or impacted wisdom teeth must be barred from the decision. But those even in the early stages of terminal or degenerative ailments clearly qualify: cancer patients, for example, and those afflicted with Lou Gehrig's disease should be free to determine the criterion, fulfillment of which is an adequate sign that they are dead. We should, in other words, leave it in their hands to decide where in the natural process of disintegration life no longer serves their human purposes. The time at which heart, lung, or brain ceases to function is not, then, for those who decide otherwise, the moment of their death; in their cases, these customary signs serve only as the physical confirmation of what took place before.

WHERE AUTONOMY IS IMPOSSIBLE

The individual-centered multiple criterion approach is obviously inapplicable, among others, to microcephalics, accident victims without living wills, and those in a persistent vegetative state. This third view focuses on the good death, and for people in such unfortunate circumstances the good death is simply not possible. In their case, the autonomy principle cannot be applied because they are not in a position to will anything or to act on their desires. For them, all we can hope for is a humane decision institutionally made.

The best place to draw the line of death in such cases is the cessation or, in some cases, the irreversible cessation of neocortical function. Our ability to sustain bodily processes beyond the point where they are of any conceivable personal benefit has generated a desperate need to distinguish between biological and human life. In view of the success of medicine and the sciences, we must make extraordinary efforts to remember that biological processes serve only as the support and substratum of human existence. When we unalterably lose the ability to will and to do, to think and to hope, to feel and love, we have ceased existence as human beings. The only humane course then is to declare us dead and to treat us accordingly. If the diagnosis is careful and accurate, we need have no fear that this harms anyone: once the human person is gone, in the faltering body there is no one there.[1]

Vanderbilt University
Nashville, Tennessee

NOTE

[1] I wish to express my thanks to John Flexner, M.D., for helpful discussions relating to the topic of this paper.

BIBLIOGRAPHY

1. President's Commission for the Study of Ethical Problems in Medicine and Biomedical and Behavioral Research: 1981, *Defining Death*, U.S. Government Printing Office, Washington, D.C., pp. 28, 77.

STUART F. SPICKER

PERSON PERCEPTION AND THE DEATH OF THE PERSON:
A NEW ROLE FOR HEALTH PROFESSIONALS IN CASES OF BRAIN DEATH

I. HUMANE TREATMENT AND THE TREATMENT OF HUMANS

Today's hospitals no longer contain only beds – the "staffed bed capacity" which affects the calculation of the hospital's revenues and expenditures – for, as Professor John Lachs remarked over a decade ago, "gardens . . . flourish in our major hospitals. . . . [They contain] thousands of human vegetables we sustain on life-preserving machines without any hope of recovery" ([4], p. 839). In this *New England Journal of Medicine* essay, Lachs was concerned to distinguish *humane treatment* from *the treatment of humans*, and he wrote convincingly on the criteria we as a society should employ in order to justify – both morally and medically – the discontinuation of treatment for those without any hope for recovery. He defended "merciful euthanasia" and admonished his readers to reevaluate their apparently fundamental aim – "to keep this creature [the unconscious vegetable] breathing and growing to no end" ([4], p. 840). Notwithstanding the ever-present danger that physicians do at times err by making false-positive determinations of death, i.e., declaring the patient dead when, indeed, he is alive, Lachs urged that we judge such irreversibly comatose creatures as no longer human. Based on these remarks, then, it seems accurate and fair to conclude that by "human" here Lachs is referring to the humanity or *personhood* of such beings, not merely to their metabolic, organic life, or "bios" – which is why he correctly refers to such metabolizing beings as mere "creatures". Nevertheless, whatever empirical tests could, with great accuracy, enable physicians to determine that the humanity of such comatose patients was non-existent would thereby serve to medically and morally justify discontinuing medical treatment, and constitute humane action on the part of the moral agents who had these patients under their care. In short, withholding or withdrawing medical treatment from non-humans that only *appear* human is the most humane medical treatment in the end. Such was Lachs's general thesis in 1976.

253

Richard M. Zaner (ed.), Death: Beyond Whole-Brain Criteria, pp. 253–265.
© 1988 *by Kluwer Academic Publishers.*

II. A REPLY TO JOHN LACHS

To-day, over a decade later, in "The Element of Choice in Criteria of Death" [5] (which precedes my remarks), Lachs has *apparently* modified his earlier position. I turn to the difficulties his current view presents:

First, Lachs fails to clarify whether brain function (partial or total) is a logical or natural necessity, and/or a logical or natural sufficiency for sustaining human personal life [1]. Furthermore, Lachs fails to strictly observe that, since there is no empirical or mechanical functional equivalent to substitute for the human brain at this time, (1) the death of the brain is a *naturally* sufficient condition for the loss of consciousness and other human qualities, and (2) the normal functioning of the neocortical brain is a *naturally* necessary condition for personal existence, as Puccetti has argued in this volume [9].

Second, Lachs trades inappropriately on various senses of 'being dead' and 'being alive'. Which sense of 'life' (and for that matter 'death') is meant is not always unambiguous. Lachs tends to vacillate between the notion of 'life' as a minimum *bios* (biological or metabolic existence) and life as human personal existence. When he says "death is a part of life" ([5], p. 248) he fails to distinguish (1) the process of *dying over time* (the 'good death') from (2) *being dead*. The more important problems, however, are revealed in Lachs's other claims:

(1) we display naïveté in our "realistic view" or ontology ([5], p. 239);

(2) we have failed to realize that death is a "biologically based social status" ([5], p. 239); and

(3) an appeal to cross-cultural anthropological evidence does *not* reveal that the temporal location of death occurs when all biological activity, down to the cellular level, has ceased, or when putrefaction sets in. That is, temporally locating death is a "social choice" ([5], p. 239).

(1) What has led Lachs to assert that a realist ontology is naïve here? Part of the answer seems to lie in his view of the meaning, significance, and conceptual *definition* of death in contrast to the empirical *criteria* for its determination. But as G. J. Agich and R. P. Jones have carefully observed: "Providing a definition of death is primarily a philosophical task and consists in saying what are the essential characteristic(s) of

death. Specifying a criterion (or criteria) is primarily a biomedical and clinical task that consists in saying what conditions must be met for the definition of death to obtain. A definition of death, then, must say what the term death means; it does not involve indicating criteria" ([1], pp. 389–390).

The question of determining death according to brain criteria, then, is primarily an empirical and mundane matter, but it is important to note that biological criteria are also derived from new conceptual understandings.

(2) Lachs remarks that "death is a biologically based social status" ([5], p. 239), and thus he affirms "the importance of biological processes" but not their "all-importance" ([5], p. 242). For the determination that 'X is dead' is a true sentence is for Lachs a "value" matter reflecting the "vital role social decisions play" ([5], p. 243). On his view, death may frequently take place *before* one is organically dead, or even brain dead. For, Lachs adds, "The time at which heart, lung, or brain ceases to function is not, then, for those who decide otherwise, the moment of their death; in their cases, these customary signs serve only as the physical confirmation of what took place before" ([5], p. 250).

Clearly and emphatically, Lachs construes the actual death of a person to be primarily a *social* determination, since death is not an exclusively physical event but a "physico-social one" ([5], p. 236). In construing biological criteria for determining that X is dead as a mere "interventionist value" ([5], p. 238) of our scientifically dominated culture, Lachs paradoxically reduces a critical *biological* moment to mere arbitrariness. But it is simply not true, as he maintains, that death is essentially a "normative" decision ([5], p. 238) and "can be situated at a number of places along the continuum of organic decline" ([5], p. 239), evidenced by cross-cultural reports.

(3) Thus, one need not reject Lachs's appeal to anthropological evidence, which reveals, for example, that Eskimo culture has sent its very elderly, frail, and burdensome members out beyond the borders of its protection (in blizzard conditions or into the night) and in so doing has sent them *to* their deaths. In short, this specific cultural behavior only serves to *make* the point that death of the frail elderly Eskimo, in the strict sense, did not take place until some time *after* these infirm persons had left the protection of the community – though it does make some sense to say (precisely because they were no longer members of

the community in any functional sense) that they were *socially* dead. A simple thought experiment, however, suffices to challenge Lachs's stress on community belief and his heavy emphasis on "social decisions", wherein he expects us to believe that any sane person could ignore X's bodily putrefaction and still honestly believe that X was alive:

We can imagine members of another culture coming upon these frail and burdensome elderly; they decide to "rescue" them from the blizzard (although they may not wish to be rescued). That is, they could be given assistance and transported to another culture's care system where they would be free to live out their remaining days. Notwithstanding the fact that this action by another culture could be judged a rather cruel act, interfering without consent in another culture's deeply held belief system, still my point remains: Social death is not death in any sense that would allow it to override cross-cultural interpretations of *organically dead* persons. Social or existential death is clearly *derivative*. Lachs himself opens his essay with the remark that "we can readily distinguish living persons from decomposing bodies" ([5], p. 233).

Though culturally important, these Eskimo elders are not dead in fact, and everyone knows it – including the Eskimos themselves. The pressing issue is not how rich or paltry is the social meaning of one's death, but whether or not one is indeed functionally dead. Whatever interest one may have in a fully described phenomenological account of the meaning of human death, that account must not avoid confrontation with the fully described condition of *being dead* and the essential organic phenomena that are necessary conditions for that status (here I am referring to the failure of anabolic processes in the various organ systems as they affect our living and/or dying human bodies).

The tendency on Lachs's part to label anyone who argues for the importance of physical, organic, or generally biological phenomena in determining the meaning and criterion of death "a naïve realist", or a misguided advocate of a "realist ontology" is surely unwarranted. We need not muddle the issue either philosophically or empirically: Given the importance of organic and especially catabolic processes, it is surely not a sufficient condition for claiming that 'X is dead' that X can no longer perform a given set of socially important activities. Lachs adds: "there has to be an appropriate biological base for the choice" of when X is dead ([5], p. 250). What should serve as conclusive evidence for the truth of the claim that 'X is dead' is the final and irreversible absence of those brain functions that constitute the naturally necessary condition of humanhood. Lachs himself concludes by rejecting the more ancient view that humans are "intimately connected with or inseparable from all of their integrated bodily acts" ([5], p. 237). For Lachs, in the end, tends to remain consistent with his view of 1976 [4]. He concludes, in choosing between neocortical and whole-brain criteria, that "the neocortical is vastly preferable" to the whole-brain criterion, and then he speaks directly to the "unwisdom of the whole-brain view" ([5], p. 247). Alas, in his concluding paragraph he says: "Our ability to sustain bodily processes beyond the point where they are of any conceivable personal benefit has generated a desperate need to distinguish between biological and human [read 'personal'] life." These remarks serve to reaffirm Lachs's position of 1976 [4]. Nevertheless, Lachs unjustifiably concludes that medicine's insistence on the *irreversibility* of critical, vital organic processes (whether heartbeat, total, or even neocortical brain function) reveals no more than medicine's tacit normative or interventionist values. We should, instead, turn Lachs's phrase on its head: *Death is a socially interpreted biological status, not a biologically based social status*, which Lachs himself admits is "less than absolute" ([5], p. 240).

What Lachs fails to stress (until the very end of his essay) is that the criterion of irreversible and permanent cessation of brain function has been adopted by contemporary medicine in order to *avoid* false-positive determinations of death – judging that X is dead when 'X is dead' is false. After all, there are always going to be difficult cases where diagnosing death with certainty is problematic, and where physicians can ill-afford declaring living persons dead persons. The empirical tests employed to determine irreversible and permanent cessation of neocortical brain function is precisely such a safety standard, established to eliminate virtually all false-positives, and is not a surreptitious insinuation of normative values into the meaning of 'X is dead.'

III. THE EMPIRICAL DETERMINATION OF DEATH

Doctors Fred Plum and Jerome B. Posner, in *The Diagnosis of Stupor and Coma*, clarify the notion of coma – generally describing it as the opposite of the state of self-awareness and the environment. But they hasten to observe that there are "a variety of altered states of consciousness" ([7], p. 1) to which a bewildering number of terms have been applied, like Roland Puccetti's use of E. Kretschmer's little-used term "apallic syndrome" [9] [*das appallische Syndrom*], first coined in 1940 ([7], p. 3, 8). This term was used to describe the behavior that accompanies the diffuse bilateral degeneration of the cerebral cortex that sometimes follows anoxic head injury, encephalitis, and a multitude of disease states, and was later used to describe patients suffering from absent neocortical function but with relatively intact brainstem function.

Hence in the context of this volume, *Death: Beyond Whole-Brain Criteria*, the contributors have not been addressing coma as such, defined as "a state of unarousable psychologic unresponsiveness in which the subjects lie with eyes closed" ([7], p. 5), but rather "neocortical death", "the apallic syndrome", "cerebral death", "akinetic mutism", or "coma vigil", where there is irreversible brain damage, but where the damage does not affect the entire brain, and patients may survive "for prolonged periods (sometimes years) without ever recovering any outward manifestation of higher mental activity" ([7], p. 6).

It should, however, be mentioned that the use of the terms noticed above presumes a greater knowledge of the associated morphological lesion than often turns out to be present at autopsy, and hence Plum and Posner prefer the term "vegetative state". They prefer it (though I do not) because "physicians and laymen both immediately understand its connotations and it has greater dignity than many of the terms sometimes applied vulgarly to the hopelessly brain damaged" ([7], p. 6). However, one can with a high degree of certainty say that large cerebral lesions or injuries damage cognitive function roughly in proportion to the amount of tissue lost and almost irrespective of what part of the hemispheres sustain the structural injury ([7], p. 16). At this point, however, it is important to mention one critical objection to the adoption of a "higher brain" standard, registered by Joanne Lynn, M.D.: "The tests available for measuring 'higher brain' death are unwieldy for use in the law because they usually involve many months of observation with only gradual accumulation of sufficient evidence to prognosticate reliably that no recovery will occur" ([6], p. 265).

Some five years have passed since Dr. Lynn worried over the ability of clinical medicine to acquire "sufficient evidence" to make accurate assessments of the brain activity of coma patients. But even by 1985, with the publication of the second edition of Plum and Posner's textbook on stupor and coma [7], it has become apparent that the empirical determination of neocortical brain necrosis is highly reliable. To be

sure, no single laboratory test or measure may prove decisive in prog-
nosticating the degree of neurologic damage in traumatic coma, or even
diagnosing brain death, but clinical signs can accurately supplement
these findings and "make the conditions obvious" ([7], p. 334). Moreover, with the advent of CT scanning in cases where intracerebral
hematoma is detected, the determination of brain necrosis no longer
adversely affects prognosis, and recent advances in determining by
scanning the degree of loss of brain tissue can assist in the localization of
hemispheric versus brain stem injury ([7], p. 334). [Here I am forced to
curtail discussion of the various pitfalls in the clinical diagnosis of
patients whose brains are *partially* necrotic ([7], pp. 324–325).]

IV. THE RATIONALE BEHIND BRAIN DEATH CRITERIA

Our contemporary world has now readdressed the concept of death and
the need for modern rather than traditional criteria for determining the
death of a person's brain in order to declare, without error, the death of
that person. It has done so for at least three reasons: (1) transplant
programs ethically conducted require the donation of healthy organs for
success, and the early diagnosis of a patient's death by means of
unassailable brain death criteria, before circulation fails, permits the
legal salvage of such organs [3]; (2) but even if availability of organs for
transplants were not necessary, still, medicine's present ability to pro-
long the life of "neomorts" (mechanically sustained brain-dead bodies)
results in futile and expensive treatment of non-persons, not to mention
the extended suffering of families and members of medical staffs –
"negativities" already shown to be inhumane by Lachs and others. The
other side is the need for new and accurately determinable criteria of
death in order to avoid false-positive determinations of death by well-
meaning physicians, when the patient should not be judged hopeless
and a reasonable expectation for recovery is indeed likely. Finally, (3)
critical care is expensive, and our society has declared virtual war on
inefficient and unjustified treatment, which requires a clearly articulated
policy of treating only those select patients who are most likely to
benefit from intensive care, and where the policy is equally clear that
medical units are not required to treat individuals who can never be
expected to recover neocortical function.

   Should such a policy be established, it will be equally important to
observe the following principle: Life support systems shall never be

discontinued in cases of brain necrosis unless the full history of medical findings determines unequivocally the degree of the brain injury and its irreversibility. Fortunately, this does not always require the use of confirmatory clinical tests other than thorough neurologic examination ([7], pp. 330–334).

Any future attempt to establish new policies toward the treatment of patients with neocortical death, however, cannot succeed in the climate which presently exists in American society. We should recall that not only the President's Commission for the Study of Ethical Problems in Medicine and Biomedical and Behavioral Research, but also the Law Reform Commission of Canada, have promulgated the view that for a declaration of death (based on the diagnosis of death of the brain) the "entire" or "total" brain, including the brainstem, must cease functioning, irreversibly. This standard, of course, has some virtues, for if the entire brain, including the brainstem, is dead, then heart beat can persist for only a few weeks or less, and these patients eventually fall in a class similar to patients with irreversible cardiac arrest.

The outcome of the U.S. President's and Canada's Law Reform Commission's reports was in effect to articulate what they perceived to be public opinion, and to do so prudently, if not on the basis of sound philosophical argument. Aside from the President's Commission's concern, that any standard other than whole-brain death would "wreak havoc with the legal system and the orderly functioning of the society" ([6], p. 265), the objection to instituting a partial or neocortical brain death standard is well expressed by Dr. Lynn, who also worries over the practical aspects of this problem: "The 'higher brain' formulation does not comport with the deeply held feelings of most people regarding the treatment of dead persons" ([6], p. 265). She echos the position taken by the Commission, which is deliberately conservative: "On a matter so fundamental to a society's sense of itself – touching deeply held personal and religious beliefs – and so final for the individual involved, one would desire much greater consensus than now exists before taking the major step of radically revising the concept of death" ([8], p. 41). Hence in cases like that of Karen Ann Quinlan, for example, who had suffered neocortical brain death but not death of the entire brainstem, Dr. Lynn argues that it seemed to be "a natural use of terms and an adequate characterization" to state that Karen Quinlan's body, until the time of total brain necrosis, was that of a living person. . . .

Is it not time to challenge this so-called "natural" use of language, as well as the view expressed by the President's Commission, where it is still accepted parlance to refer to a neocortically-dead body as the body of a "living person"?

Here we should be reminded of the contribution of S. J. Youngner and E. T. Bartlett [10]. These authors urge us to begin to work to "change the public's perception of the behaviors" that should follow designating patients like Ms. Quinlan as neocortically brain dead. At the very least, short of altering the public's perception and language, health care professionals should not be obliged to treat "neomortic life". Having said this, of course, much more must be added by way of qualification, since Dr. Lynn's concern is not easily dismissed, and warrants careful review ([6], p. 265). That is, any attempt to alter the public's perception of and language about the "behavior" of neocortically dead bodies (which the public persists in seeing as the bodies of persons) will be no simple achievement. As Lachs has remarked: ". . . it is still difficult for us to believe that individuals who appear physically intact and may even breathe on their own could, nevertheless, be irremediably dead" ([5], p. 246). And as Marx Wartofsky has observed, the residuum [neomort] has itself human meaning; there is, so to speak, a "residual personality" to consider, as well as the prior person's humanity in the residual memory of the living. [This follows from the ontological truth that *the dead are not nothing*, whereas all mythological figures are, strictly speaking, nothing at all, and they could not come to exist, either; their names only reflect our pretense that they share some mode of existence. By the way, here I disagree with Lachs's claim that "the job of philosophers is not so much to resolve thorny ontological problems about the personhood of human beings, as to focus public discussion and political decisionmaking by the critical examination of our beliefs and of the consequences of the values we embrace" ([5], p. 245).]

V. PERSON PERCEPTION AND THE APPRAISAL OF PERSONS

In the 1950's, gestalt psychologists T. Dembo and Beatrice A. Wright undertook research to determine how persons with negative attributes, such as disabilities or disfigurements, are perceived by others who do not suffer from these negative attributes. They set about to study the rational and irrational cognitive processes which lead us to generalize that other inferred characteristics of such persons are equally negative.

They subsequently learned that negative information about handicapped persons generally carries more weight with non-handicapped
persons than positive information about such persons. In short, we tend
to perceive or appraise handicapped persons as more handicapped than
they are. This led the researchers to postulate two psychic frameworks
or vastly different orientations, which they labelled the "coping" and
the "succumbing". Once their research was completed, it was clear that
even health professionals needed to work to modify their perception of
the handicapped from the "succumbing" to the "coping", without being
frightened into the future mode of person perception when working
with the handicapped and the disfigured. In short, the researchers
emphasized the results of their work in a maxim: "a person is not
equivalent to an impairment" ([11], p. 23). Having worked extensively
with the phenomenon of *spread* – seeing those persons with negative
attributes as worse off than in fact they are – Wright concluded that,
"Despite continuing efforts to enhance human potential, however, the
probability remains that negative spread will continue to exist as a
social-psychological obstacle, although hopefully less pervasively so."
She concludes, "This follows because not all conditions underlying the
phenomena of negative spread can be eliminated by changing beliefs, or
values, or economic conditions" ([11], p. 23).

   In the context of our perception of "persons" who suffer neocortical
brain death, we can perhaps suggest reapplying the research of Wright
and Dembo. We can stand it on its head, so to speak: In employing the
"succumbing" and "coping" modes of person perception, our problem
is one of changing the public's perception of the neocortically-dead
person from a being that is capable of coping, doing, acting, and of
experiencing satisfaction, to a being that has succumbed, can no longer
act, and must remain passive. In so doing, we would be attempting to
alter the public's future perception of neomorts as beings for whom
hope for recovery is *no longer* possible, due to permanent loss of self,
although there would still remain human meaning in the "residual
personality" who once lived in that "earthly form". A psychic ambiguity
would be revealed in (1) the residuum which lies in bed approached with
respect, while (2) that "same" being is perceived as no longer the living
presence of a person. In the short run, neocortically brain-dead bodies,
artificially sustained by respirators or ventilators, will still appear alive,
due to moving chest, pulsing blood vessels and bodily warmth. To
suggest that we bury such beings while they breathe and have heartbeat

is, to-day, esthetically unacceptable, if not morally repulsive, disrespect-
ful, and a desecration in the minds of many. Given our imprinted
present attitude, who will accept the mandate to terminate the final vital
processes of this *apparently* living but actually neocortically-dead patient?

Perhaps this special role should be assumed by psychiatric medicine.
After all, although it is reasonable to suggest that all health care
professionals – especially nurses and physicians – begin the slow process
of acquiring a new perceptual attitude toward the neocortically-dead
patient, perceiving him as truly dead, it may not be inappropriate to
suggest that since psychiatrists are more expert in matters of cognition
and perception, that they take the lead in working to alter the public's
perception of neomortic life; that they assist us, by working with
patients' families and others, to *see* such "patients" as having succumbed to death, no longer able to recuperate. In short, we shall once
again have to learn to see – to apprehend in one's very seeing that the
person of the patient has evanesced, that the patient of a moment ago in
his earthly form is gone, and can therefore no longer cope in spite of our
best efforts to affect recovery.

At the very least this means that we might, but only in time, come to
intervene to end clinically diagnosed neomortic life, while all the time
avoiding any desecration of the residuum; for one must not do violence
to the body, but rather acknowledge in "secular piety" the irreversible
loss of the patient's earthly form ([2], p. 130). This will require a slow
and evolutionary change in the public's perception and, subsequently, a
change of its ordinary ways of speaking. Only in time will it become
clear that although the *living bodily presence* of a patient with minimal
neocortical function entails the living presence of that person, *the body's
living presence* only refers to the neomort, the neocortically-dead person, to which it is inappropriate to attribute pain and suffering. Indeed,
it is difficult to imagine doing harm to patients with apallic syndrome,
other than doing violence to the neomort by using the residual body as a
mere means.

In addition, we shall have to come to distinguish clearly between the
death of *the organism as a whole* (the intrinsic unity annihilated with
neocortical death) and *the death of the whole organism* (each and every
part of the organism); the former is an event, whereas the latter occurs
over time and is thus a process of extended duration. Since clinical
medicine can now accurately determine the death of the organism as a

whole from a diagnosis of neocortical death, organ transplantation programs, for example, are morally permissible. The death of the organism as a whole, then, is a sufficient condition for declaring the death of the patient.

VI. CONCLUDING WORD

It has taken some time to learn that persons are neither mindless bodies nor disembodied minds; now we shall have to come to see that brain-stems, as well as hearts and other peripheral organs, may clearly outlive their owners. This "disexistential state" (if I may call it so) is precisely the state of neocortically-dead bodies evidenced by the continuous growth of various cell types: hair, fingernails, etc. To defer the recognition and declaration of personal death until *all* biological life or the entire brain becomes necrotic is to invoke a criterion of death which calls for the cessation of *all* physiological activity. Such a criterion, I suggest, is not coherent with our frequently professed commitment (or what Lachs would call *value*) to respect the life of persons.

The University of Connecticut School of Medicine
Farmington, Connecticut

BIBLIOGRAPHY

1. Agich, G. J. and Jones, R. P.: 1985, 'The Logical Status of Brain Death Criteria', *The Journal of Medicine and Philosophy* **10** (4), 386–395.
2. Jonas, H.: 1974, 'Against the Stream: Comments on the Definition and Redefinition of Death', in Jonas, H., *Philosophical Essays,* Prentice-Hall, Inc., Englewood Cliffs, New Jersey, pp. 132–140.
3. Joul-Jensen, P.: 1970, *Criteria of Brain Death: Selection of Donors for Transplantation,* (trans.) A. Rousing, Munksgaard, Copenhagen, Denmark.
4. Lachs, J.: 1976, 'Humane Treatment and the Treatment of Humans', *The New England Journal of Medicine* **291** (15) (April 8), 838–840.
5. Lachs, J.: 1988, 'The Element of Choice in Criteria of Death', in this volume, pp. 233–251.
6. Lynn, J.: 1983, 'The Determination of Death', *Annals of Internal Medicine* **99** (2), 264–266.
7. Plum, F. and Posner, J. B.: 1985, *The Diagnosis of Stupor and Coma*, 3rd ed., F. A. Davis Co., Philadelphia, Penn.
8. President's Commission for the Study of Ethical Problems in Medicine and Biomedi-

cal and Behavioral Research: 1981, *Defining Death*, U.S. Government Printing Office, Washington, D.C.

9. Puccetti, R.: 1987, 'Does Anyone Survive Neocortical Death?', in this volume, pp. 75–90.

10. Youngner, S. J. and Bartlett, E. T.: 1983, 'Human Death and High Technology: The Failure of the Whole-Brain Formulations', *Annals of Internal Medicine* **99** (2), 252–258.

11. Wright, B. A.: 1979, 'Atypical Physique and the Appraisal of Persons', *Connecticut Medicine* **43** (10), 19–24.

NOTES ON CONTRIBUTORS

Edward T. Bartlett, Ph.D., is Associate Professor of Philosophy, Cleveland State University, Cleveland, Ohio.

Alexander M. Capron, L.L.D., is Norman Topping Professor of Law, Medicine, and Public Policy, The Law Center, University of Southern California, Los Angeles, California.

H. Tristram Engelhardt, Jr., Ph.D., M.D., is Professor, Center for Ethics, Medicine, and Public Issues, Baylor College of Medicine, Houston, Texas.

John Lachs, Ph.D., is Professor of Philosophy, Vanderbilt University, Nashville, Tennessee.

Martin S. Pernick, Ph.D., is Associate Professor, Department of History, The University of Michigan, Ann Arbor, Michigan.

Roland Puccetti, Ph.D., is Professor, Department of Philosophy, Dalhousie University, Halifax, Nova Scotia, Canada.

Charles E. Scott, Ph.D., is Professor and Chairman of Philosophy, Vanderbilt University, Nashville, Tennessee.

David R. Smith, J.D., is Assistant Professor of Law, Vanderbilt University, Nashville, Tennessee.

Stuart F. Spicker, Ph.D., is Professor of Community Medicine and Health Care (Philosophy), School of Medicine, University of Connecticut Health Center, Farmington, Connecticut.

Robert M. Veatch, Ph.D., is Senior Research Scholar. The Kennedy Institute, Center for Bioethics, Georgetown University, Washington, D.C.

Marx W. Wartofsky, Ph.D., is City University Distinguished Professor of Philosophy, Baruch College, City University of New York, New York.

Patricia D. White, J.D., is Associate Professor of Law, Georgetown University Law Center, Washingtown, D.C.

Stuart J. Youngner, M.D., is Associate Professor of Psychiatry and Medicine, School of Medicine, Case Western Reserve University, Cleveland, Ohio.

Richard M. Zaner, Ph.D., is Ann Geddes Stahlman Professor of Medical Ethics, Department of Medicine, Vanderbilt University, Nashville, Tennessee.

INDEX

The Philosophy and Medicine Book Series

Editors

H. Tristram Engelhardt, Jr. and Stuart F. Spicker